TOWARD SUSTAINABILITY

A Plan for Collaborative
Research on Agriculture and
Natural Resource Management

Panel for Collaborative Research Support for AID's
Sustainable Agriculture and Natural Resource
Management Program

Board on Agriculture

Board on Science and Technology
for International Development

National Research Council

NATIONAL ACADEMY PRESS
Washington, D.C. 1991

This report has been prepared with funds provided by the Bureau for Science and Technology, Office of Agriculture and Office of Rural Development, U.S. Agency for International Development, under Grant No. DAN-5052-C-00-6037-00. The U.S. Agency for International Development reserves a royalty-free and nonexclusive and irrevocable right to reproduce, publish, or otherwise use and to authorize others to use the work for government purposes.

Library of Congress Catalog Card No. 91-61818
ISBN 0-309-04540-1

A limited number of copies are available from:
Board on Science and Technology
    for International Development
Office of International Affairs
National Research Council
2101 Constitution Avenue
Washington, DC 20418

Additional copies are available for sale from:
National Academy Press
2101 Constitution Avenue
Washington, DC 20418

S-378

Printed in the United States of America

PANEL FOR COLLABORATIVE RESEARCH SUPPORT FOR AID'S SUSTAINABLE AGRICULTURE AND NATURAL RESOURCE MANAGEMENT PROGRAM

LOWELL HARDIN, *Chairman,* Purdue University
JOHN AXTELL, Purdue University
HECTOR BARRETO, Centro Internacional de Mejoramiento de Maíz y Trigo, Guatemala
BARBARA BRAMBLE, National Wildlife Federation
PIERRE CROSSON, Resources for the Future
CLIVE EDWARDS, Ohio State University
RICHARD HARWOOD, Michigan State University
G. EDWARD SCHUH, Hubert H. Humphrey Institute of Public Affairs, University of Minnesota
G. K. VEERESH, University of Agricultural Sciences, India
ROBERT WAGNER, Phosphate and Potash Institute (Retired)

Ex Officio Members

PATRICIA BARNES-MCCONNELL, Collaborative Research Support Program, Michigan State University
LEONARD BERRY, Florida Atlantic University
PEDRO SANCHEZ, North Carolina State University
JAN VAN SCHILFGAARDE, U.S. Department of Agriculture, Ft. Collins, Colorado

Staff

MICHAEL MCD. DOW, *Study Director*
JAY DAVENPORT, *Senior Project Officer*
CURT MEINE, *Staff Associate*
NEAL BRANDES, *Study Assistant*
NANCY NACHBAR, *Program Assistant*

iv

Preface

In response to growing support for sustainable international development strategies, the U.S. Congress has recommended that the Agency for International Development (AID) create a new Collaborative Research Support Program (CRSP) that focuses on the research needs of sustainable agriculture and natural resource management. The Office of Agriculture in AID's Bureau for Science and Technology subsequently asked that the National Research Council's Board on Agriculture (BA) and Board on Science and Technology for International Development (BOSTID) undertake planning for the new CRSP.

Collaborative research support programs were created under Title XII of the International Development and Food Assistance Act of 1975, which supports long-term agricultural research of benefit to developing countries and the United States. These programs are the primary mechanisms through which U.S. universities conduct such research. Currently eight CRSPs are conducting research on several important crops, livestock, soils, fisheries, aquaculture, and human nutrition.

The charge to the National Research Council's Panel for Collaborative Research Support for AID's Sustainable Agriculture and Natural Resource Management Program was to: (1) recommend a design for the new CRSP; (2) help AID define research priorities for the new CRSP; and (3) suggest management arrangements for administering the CRSP that will enable it to draw on and contribute to all of AID's agricultural, environmental, and rural development activities. Officials of AID requested that the panel, in carrying out its charge, try to define a process by which knowledge from all relevant AID-supported research, development, and training programs could be integrated and applied in the effort to advance profitable farming systems that improve local conditions while contributing to broader environmental goals.

The panel is one of three units established at AID's request to assist the Office of Agriculture in reviewing its projects on sustainable agriculture and natural resource management. The Committee for a Study on Sustainable Agriculture and the Environment in the Humid Tropics is studying successful approaches to sustainable agriculture in the humid tropics. Its activities are managed jointly by BA and BOSTID. The Committee International Soil and Water Research and Development is assessing the needs and priorities in soil and water management for developing countries. Its activities are managed jointly by BOSTID and the Water Science and Technology Board.

The Panel for Collaborative Research Support for AID's Sustainable Agriculture and Natural Resource Management Program has focused on the need to promote integrated, multidisciplinary research across agroecological zones, among departments and institutions of U.S. universities, and in collaboration with other institutions, research institutes, national agricultural research systems, and the international agricultural research centers. Its principal objectives have been to foster a truly collaborative and participatory approach to the design of research and to involve the ultimate beneficiary of the research: the small-scale farmer and rural and urban poor in developing countries. From its inception, the panel has emphasized the need to draw on and actively engage in-country expertise and indigenous knowledge and practices in meeting its objective.

At an organizational meeting in July 1990, participants stressed the fact that research under the new CRSP must focus on on-farm methodologies that effectively integrate the agronomic, biological, ecological, cultural, and socioeconomic factors that govern the performance and sustainability of agroecosystems. Only such integrated research can fill the critical gaps in scientific understanding of the foundation and functioning of sustainable agricultural systems. Of particular importance in this regard are the following:

• Conservation of soil and water resources and the impact on fertility of the soil's physical and biological characteristics, processes, and cycles;
• Cultural practices for improving soil fertility, controlling erosion, and maximizing biological production potential (for example, tillage methods, crop residue management, irrigation, alley cropping, and agroforestry);
• Integrated pest management systems, both pre- and postharvest;
• Indigenous practices and uses of germplasm and the economic and cultural consequences of biodiversity loss and preservation;
• The consequences of converting forest and savannah lands into range for cattle production;
• Institutional arrangements—local, national, and international—involving education, trade, finance, and prices;

• Common issues related to property resource management, land tenure, and other public policies; and

• The impact of policy incentives or disincentives on the production of cash crops for export or food crops for local consumption.

The development of research methodologies to address these key gaps in knowledge is a formidable task. The further implementation of the necessary research to fill these gaps will require an enormous commitment of resources over an extended time. Participants in the organizational meeting agreed that the new CRSP should not be restricted to, but should concentrate on, the more fragile agroecosystems in targeting its initial investments for maximum effect. They also noted the need for an open planning process for the CRSP. To this end, the panel together with invited participants from the land-grant colleges and universities and other interested organizations— more than 120 people—convened in November 1990 for an open forum on international sustainable agriculture and natural resource management. At the day-long forum, invited speakers and other participants reviewed the CRSP record and the experience of collaborative international agricultural research at U.S. universities. During 3 days of intensive follow-up discussions, participants discussed research priorities and suggested guidelines for establishing and managing a program to encourage research on sustainability, agriculture, and natural resources in U.S. institutions and their developing country counterparts.

The panel met twice after the November forum. This report summarizes the findings from the forum and the subsequent panel discussions. An executive summary provides a synopsis of the rationale and principal recommendations for the new Collaborative Research Support Program on Sustainable Agriculture and Natural Resource Management. The panel's findings and specific recommendations are then presented in greater detail in the main body of the report. The papers presented at the open forum and the discussions that followed generated several significant statements on agroecosystem research and management. These are included as appendixes. A concurrent subpanel was convened to summarize and provide guidance to AID on activities involving integrated pest management, an area of particular importance to sustainability. The discussions of the subpanel will be published in a separate report in late 1991.

The panel has tried to accommodate as faithfully as possible the many viewpoints germane to this topic. The panelists and participants in the November forum, though diverse, were in fact in welcome accord on one principal point: the need for research to focus on the integration of the social and natural sciences in progressing toward sustainability. Not all participants would agree on the means of accomplishing this challenging task. Further, the report does not deal in any depth with population policy and family planning concerns, which are important factors in the sustainability formula. Nonetheless, within the scope of this report, the broad consensus regarding the nature of the scientific and managerial challenge bodes well for the future. In particular, the challenge of bringing together the varied disciplines, with their different traditions, approaches, and languages, must be met to gain a better understanding of the nature of sustainability.

Members
Panel for Collaborative Research
Support for AID's Sustainable
Agriculture and Natural Resource
Management Program

Acknowledgments

As with all endeavors that try to bring different perspectives together and distill large amounts of technical information into a coherent form, this effort has been a challenging one. The panel deeply appreciates the extensive advice it received in the short time available for completion of this report. The panel is entirely responsible for any shortcomings of the report.

Several people deserve special thanks: those who participated in and, in many cases, prepared written papers for the forum and subsequent workshop, and who later commented on the draft report; others who were unable to attend the meeting but who reviewed and offered comments on the draft; and Thurman Grove, for his substantive assistance as liaison at the Agency for International Development.

We would also like to acknowledge the intellectual contributions of Charles Benbrook and Charles B. McCants. Invaluable assistance was provided by Jay Dorsey, Chris Elfring, Patricia A. Harrington, Mary Francis Schlichter, and Lynn Wolter.

Contents

Executive Summary

Many agricultural and natural resource management practices are increasingly implicated in environmental deterioration around the world. The symptoms include soil erosion and other forms of soil degradation, deforestation and desertification, declining water quality and availability, the disruption of hydrogeological cycles, and the loss of biological diversity. Land use practices may also be affecting regional and global climatic patterns. These interrelated phenomena, in turn, can lead to losses in agricultural productivity at local and regional levels, and they raise concerns about food security, food quality, public health, and other long-term development issues.

The symptoms and human costs of environmental deterioration are evident everywhere to varying degrees, but they are of special concern in the developing nations of the tropics, where soils are often shallow, highly weathered, low in fertility, and easily eroded; where agricultural ecosystems are subject to a greater number and variety of diseases, weeds, and other pests; where biological diversity is so remarkably rich—and at greatest risk; and where economic constraints and development needs are most pressing.

The size of the human population is expected to increase by 1 billion people—the equivalent of an additional China—each decade well into the next century. Most of this growth will occur in developing nations, where the limits of available arable land are being reached. In light of these expectations, environmental quality and economic development can no longer be considered separately.

THE CONCEPT OF SUSTAINABLE AGRICULTURE

Sustainable agriculture is a relatively recent response to these environmental and economic concerns. Early discussions of the concept stressed the importance of the renewal capacity of agricultural ecosystems and claimed that many conventional agricultural practices were detrimental to this capacity. Out of further discussion has emerged an approach to agriculture that incorporates the principles of ecology by emphasizing interactions among and within all the components of agroecosystems (including, by definition, the social and economic components).

As more individuals and organizations have begun to recognize the need for adjustments to conventional agriculture that are environmentally, socially, and economically compatible, the phrase *sustainable agriculture* has come to connote approaches to agriculture that provide for the needs of current and future generations while conserving natural resources. Indeed, a major development in the past decade has been the emerging recognition on the part of agricultural production and environmental management groups that they share common, rather than competing, goals. In this context, sustainable agriculture is often used to refer to agriculture and all its interactions with society and the greater environment; as such, it can be considered a vital component of current discussions of sustainable development.

The literature offers hundreds of definitions of sustainable agriculture, virtually all of which incorporate the following characteristics: long-term maintenance of natural resources and agricultural productivity, minimal adverse environmental impacts, adequate economic returns to farmers, optimal crop production with minimized chemical inputs, satisfaction of human needs for food and income, and provision for the social needs of farm families and communities. All definitions, in other words, explicitly promote environmental, economic, and social goals in their efforts to clarify and interpret the meaning of sustainability. In addition, all definitions implicitly suggest the need to ensure flexibility within agroecosystems in order to respond effectively to stresses. These characteristics of sustainable agriculture provide a framework and suggest an agenda for the evolution of agriculture and natural resource management to meet the needs of changing societies and environments.

THE RESEARCH CHALLENGE

Fundamentally, achieving sustainable agriculture under the mounting pressure of human population growth will demand that the world's agricultural productive capacity be enhanced while its resource base is conserved. If the well-being of the world's less advantaged people is to improve in any lasting sense, long-range concerns about food security and the health of natural resources must be addressed in planning future economic and social development. Research on sustainable agriculture and natural resource management will be essential to this task. More specifically, researchers must devote greater attention to developing integrated cropping, livestock, and other production systems—and the specific farming practices within these systems—that enhance (or, at minimum, do not degrade) the structure and functioning of the broader agroecosystem. Most agricultural research focuses on single commodities, components, or disciplines within agriculture. More research is needed that approaches agriculture in an integrated, interdisciplinary manner.

The Need for a Sustainable Agriculture and Natural Resource Management Collaborative Research Support Program

The collaborative research support programs (CRSPs) of the Agency for International Development (AID) are the main mechanisms through which U.S. universities implement Title XII of the International Development and Food Assistance Act of 1975, which supports agricultural research of benefit to developing countries and the United States. To date, eight CRSPs have been established. They are focusing their research efforts on specific commodities (sorghum and millet, beans and cowpeas, and peanuts), livestock (small ruminants), soils, fisheries, aquaculture, and human nutrition. The distinguished research record of these CRSPs, and their important contributions to solving agricultural problems, are recognized worldwide.

The importance and timeliness of research into sustainable agriculture and natural resource management, and the need for integrated approaches to this research, demand that a new CRSP be implemented as soon as possible. Moreover, sustainability and agroecological considerations are so important and central to attaining development goals that they should be fundamental to planning and carrying out all the agricultural and natural resource programs that AID supports. Thus, the new CRSP should not be viewed as the only AID sustainable agriculture activity; all other AID-supported activities, including the existing CRSPs, address various aspects of sustainability, and they must continue to do so. The new CRSP should complement these existing efforts and add a critical dimension of integration as the core activity of a comprehensive Sustainable Agriculture and Natural Resource Management (SANREM) program. The program, proposed herein, should include the CRSP and related collaborative research activities funded by AID. It should serve to stimulate and support innovative, integrated systems-based collaborative research into the ecological and socioeconomic characteristics of sustainable agriculture and natural resource management within the world's major agroecosystems.

Commitment to Systems-Based Research

Across all systems, sustainability implies the securing of a durable, favorable balance of economic and environmental costs and benefits. An integrated systems approach, whether defined formally or informally, is therefore essential to all research under the proposed SANREM program. The research location should encompass a landscape or political unit of sufficient size and diversity to support studies of all the principal determinants of sustainability within the agroecosystem. To the fullest extent possible, farmers should actively participate in each phase of the research process, from initial planning and testing to technology development, dissemination, and other extension-related activities. An appropriate balance of university research station and farmer-field effort is recommended. Because considerable attention is already being given to input-intensive agroecosystems, efforts should be directed primarily, but not exclusively, to the more fragile agroecosystems.

The SANREM effort would benefit not only the developing countries in which it is conducted and to which it is directed, but also the United States, through the development of more effective research methodologies, the training of U.S. researchers, and the acquisition of results pertinent to the sustainability of U.S. agriculture and natural resources.

Commitment to Interdisciplinary Inquiry

The goal of sustainability and the scientific questions it raises are complex. Accordingly, research conducted under the SANREM program should involve natural, agricultural, and social scientists who have a commitment to interdisciplinary inquiry. This commitment must be shared by collaborating institutions and local governments if the program is to succeed.

Research should take into consideration all the basic elements involved in agricultural systems performance (including soil and water resources, tillage and cultivation methods, cropping patterns, animal husbandry, nutrient management, and pest management), but it should devote attention to additional components (such as aquaculture and farm forestry) as appropriate. Resource policies and other institutional factors play a critical role in determining the choices that farmers make and, hence, the sustainability of farming systems. Accordingly, research must also be directed to the socioeconomic and policy context within which farmers make their decisions.

Knowledge of all relevant components and their interactions is fundamental to understanding the functioning and management of agroecosystems. However, this knowledge is often inadequately integrated or lacking altogether. Greater understanding of the sustainability of agroecosystems will require that all relevant factors be researched, and that they be researched together.

Research Approach

It is not possible to prescribe here recommendations or research priorities for specific locations. The conditions conducive to sustainability in any particular agroecosystem, or at any particular site, will differ depending on the constraints, opportunities, and interrelationships among various factors at that location. However, certain factors—soil conditions, water quality and availability, biodiversity, nutrient cycling, pest pressures, cultural traditions, economic incentives, and public policy—affect all sites and agroecosystems, and together they help determine the sustainability of the system. Thus, the SANREM program should encourage an approach to research that emphasizes these cross-cutting ecological and socioeconomic concerns.

Special attention should be given to the following areas of inquiry, which are the least understood and least researched topics common to all agroecosystems. *Integrated pest management* seeks to control pre- and postharvest weeds, arthropod and vertebrate pests, and pathogens using biological and cultural techniques along with minimal levels of synthetic pesticides. *Integrated nutrient management* seeks to provide plant nutrients through the optimal use of on-farm biological resources (including manures, plant rotations, cropping patterns, and legumes) and, where necessary, purchased inputs. Integrated pest and nutrient management depend on conserving biological diversity and soil organic matter and, thus, on a sound understanding of biological processes and ecological interactions.

Greater attention should also be given to research on *integrated institutional management,* including a production economics component, to guide the complex interactions between food and fiber production and the policy, trade, and political environments. The *social, political, and institutional contexts* within which both on-farm and off-farm activities take place must also be given greater attention to identify those opportunities that can be reinforced, and those constraints that can be removed, to promote sustainability. This calls for a strong and innovative social science component in the research design that is focused on the institutional and policy conditions that influence on-farm resource management patterns. This research should address issues of gender and age, the impact of production alternatives on social structure, and ways to strengthen critical human resources, including especially local and indigenous knowledge. If the adoption of more sustainable methods and technologies should involve hardship for some local farmers, such results should be anticipated, forthrightly acknowledged, and studied with a view toward amelioration.

THE GRANT PROGRAM

Progress toward the objectives of the proposed SANREM program should be furthered through competitive research grants. (To support research activities, AID employs contracts, cooperative agreements, and grants. In this report, grant is used generically to refer to all of these mechanisms.) No single, established model exists for the successful conduct of the integrated, multidisciplinary research and development efforts that the SANREM program would require. Thus, the grant program should be designed so that maximum reliance is placed on the ingenuity of the researchers who will do the work. Innovative research design, reflecting creative approaches to the full range of sustainability issues, should be the key criterion for research sponsored under this program. Research proposals should reflect this in following the guidelines and meeting the requirements set forth below. A competitive, peer-review granting process is the most effective means of identifying research proposals that meet these criteria and requirements.

Grant Types

Three types of competitive grants should be made available under the SANREM program: research planning grants, a research core grant, and research support grants.

Research planning grants should support enhanced interdisciplinary interaction, on-site visits to potential host countries, and the development of links with cooperating institutions in the process of preparing and refining proposals for the research core grant. A maximum of six planning grants of up to $50,000 each per institution or consortium should be awarded during the initial year of the program.

A research core grant should support a long-term, full-scale interdisciplinary collaborative research program (the SANREM CRSP) on sustainable agriculture and natural resource management in one or more of the world's principal agroecosystems. It should be awarded in the second year of the program at a level of about $2.5 million annually.

Research support grants should support research of direct and immediate relevance to the goals of the SANREM program within other collaborative research programs, including existing CRSPs. Two types are recommended: type A, to be awarded by the CRSP management entity as soon as the SANREM CRSP is established; and type B, to be awarded directly by the AID Bureau for Science and Technology as soon as possible. A limited number of grants of up to $100,000 per year should be awarded for an initial 3-year period.

Institutional Participation

Research conducted under the SANREM program would demand a broad range of expertise and international experience in the natural, agricultural, and social sciences. To be successful, projects may require the involvement of organizations and institutions that are not currently Title XII program participants. All colleges and universities should be eligible to receive SANREM program funds, and subcontracts should be available to other groups with the requisite expertise, including private voluntary, nongovern-mental, and other private sector organizations. The SANREM program should capitalize on the research and development capabilities of the entire U.S. system and of diverse collaborators in developing countries. Since collaboration with host country institutions would be essential to achieving SANREM goals, subcontracts with relevant developing country entities would be encouraged.

Content of Research Proposals

In evaluating grant proposals, and thereafter in monitoring and evaluat-ing funded research, AID should require that applicants provide information and demonstrate capacities as indicated in the following list:

- description of research location and site description;
- significance of research and site;
- problem description and research methodology;
- systems-based approaches to ecological and socioeconomic research;
- capacity for interdisciplinary research;
- capacity to develop technologies and disseminate knowledge;
- collaborative arrangements among U.S. and host country institutions;
- information about researchers and other collaborators; and
- budget.

Proposals for research planning grants and the research core grant should meet the same set of requirements to the fullest degree possible. Research support grant proposals, on the other hand, should meet those requirements from among this list as necessary to augment their established research agenda.

Administrative Procedures

To achieve the grant objectives, AID should observe the following pro-cedures in administering the grant program:

• Current CRSP guidelines, with modifications as needed to meet the broader SANREM program goals, should be followed and made available to all potential applicants.

• Expanded planning grant proposals can serve as final core grant pro-posals, but core grant applicants should not be required to have applied for, or to have received, a planning grant.

• The awarding of type B research support grants should neither hinder nor promote the eligibility of the same institution for the core grant.

• All SANREM grant applicants should be required to adhere to the special concerns guidelines for research grants required by AID's Program in Science and Technology Cooperation (Agency for International Development, 1990). These guidelines, which pertain to the handling of genetic materials, pesticides, radioactive and other hazardous materials, and other concerns, should be made available to all potential applicants.

Program Timetable

In awarding the research planning grants and research support grants, and in selecting the core grant recipient and management entity, the timetable outlined in Chapter 4 (Table 4-1) should be followed.

CONCLUSION

The establishment of the proposed SANREM program, and the competitive grants it would make available, would provide focus and support for collaborative research on agricultural sustainability. Although the need for new approaches, innovative experimental designs, and integrated training in support of sustainable agriculture and natural resource management has been recognized for some time, the institutional and financial means to implement responses have been scarce. Research of the kind needed is long term and complex, requiring sustained commitment that a new collaborative research support program can provide. Although a modest step given the extent of the challenge, the establishment of the SANREM program should catalyze support from other parts of AID and from other donor agencies, and contribute directly to developing sustainable agricultural systems and natural resource management strategies.

1

Defining the Need

As concerns about environmental protection, natural resource steward-
ship, and the world's ability to feed ever-growing populations continue to
mount, the sustainability of agriculture and natural resources is emerging as
a central theme among the public and policymakers alike. The importance
given to it reflects the recognition that the quality of human life and the
quality of the environment are inextricably linked. The issues involved
transcend science. They encompass ideologies and values, ethics and aes-
thetics—the arena, in short, of public opinion and public policy. The issues
also transcend national boundaries and involve critical considerations of
intergenerational responsibility and equity.

The deepening awareness of the interdependence of agriculture, the environment, and socioeconomic conditions has called into question the sustainability of current agricultural production systems. In industrial countries,
the environmental effects of intensified production have led many to search
for ways to maintain and enhance productivity through better management
of the entire agricultural system, including changes in socioeconomic incentives and policies.

The recent National Research Council (1989a) report *Alternative Agriculture* describes the human and environmental costs of high-input production methods in the United States. Based on a growing body of research
and experience, the report examines the environmental problems that today's
widely accepted agricultural practices can cause or fail to prevent. These
include soil erosion and degradation, nonpoint source water pollution, groundwater contamination, salinization, aquifer depletion, loss of biological diversity, resistance to pesticides, and human health risks associated with
pesticide application and residues.

The report calls attention to the economic and environmental effects of reduced reliance on chemical pesticides and fertilizers, and in a series of case studies describes the experiences of farmers who have adopted alternative practices, including crop rotation, integrated pest management, and increased use of on-farm nutrient sources. These innovative farmers have taken the lead in devising and implementing new management approaches on their farms, and the case studies document the results—the successes as well as the failures—from their fields, pastures, and orchards. The report argues that research needs to be directed toward alternative practices and improvements in technology and management know-how. It also calls for research on the social, economic, institutional, and policy factors that influence the choices farmers make. Such research can contribute to the formulation of incentive programs that encourage the development and adoption of beneficial alternatives.

Many of the same forces, trends, and interdependencies described in *Alternative Agriculture* are important in other areas and agroecosystems around the world. Additional factors, especially continued rapid population growth and crushing poverty, increase the pressure on the land and accelerate the processes of environmental deterioration. They are particularly acute in developing countries, where people are unable to buy food, governments are unable to purchase food on world markets, and distribution problems hinder availability even when local supplies are adequate. As some areas exhaust their supplies of arable land, inappropriate land use practices are causing massive soil erosion, critical losses of biological diversity, and general degradation of the natural resource base. In the tropics, where these forces are especially potent, the burning of rain forests to clear land for agriculture adds to the threat of global warming. Global agriculture and resource management thus face alarming problems as the twenty-first century nears.

AGRICULTURE, ENVIRONMENT, AND DEVELOPMENT

The human population is expected to increase by 1 billion people—the equivalent of an additional China—each decade well into the next century. Most of this population growth will occur in the developing nations, placing further stress on their arable land bases. In many countries, the limited availability of arable land, combined with urban congestion, has led to spontaneous and organized migrations and the clearing of new land for agriculture. Land clearing has contributed directly to the degradation of soil, water, and other natural resources in both humid tropical and semiarid countries.

In the humid tropics, conversion of the rain forest for agriculture, timber, and large-scale ranching is accompanied by the loss of topsoil and the depletion of nutrients, especially nitrogen, through leaching of exposed soil or through volatilization by the burning of land for clearing (Lal, 1986; Pimentel et al., 1987). The loss of soil in the uplands results in degradation of inland and coastal waters and disruption of hydrogeological cycles.

The forests of the humid tropics are also the world's richest repositories of biological diversity, and deforestation threatens to drive many forest species, many not yet even identified by science, to extinction. Numerous reports (McNeely, 1988; Myers, 1980; National Science Board, 1990; Office of Technology Assessment, 1987; Wilson, 1988) document the value of biodiversity and describe the extensive and varied consequences for agriculture of reduced diversity. These consequences include losses of plant and animal species with the potential for domestication; genetic strains resistant to drought, pests, and disease; beneficial pollinators and symbionts; and pest antagonists, parasites, and predators. Destruction of the rain forests also contributes, through increased rates of biomass decomposition, burning, and oxidation of soil organic matter, to the buildup of atmospheric carbon dioxide and other greenhouse gases (Crutzen and Andrae, 1991; Houghton, 1990; Myers, 1989; U.S. Environmental Protection Agency, 1990).

In arid and semiarid areas, demands for wood, fuel, fodder, and shelter increase with the growth of populations of people and livestock. The environmental results are analogous to those affecting the tropical rain forests (National Research Council, 1984). In the Sahel, overgrazing by cattle and sheep, which in many areas have replaced browsing camels and goats, has resulted in the conversion of grasslands from deep-rooted perennial grasses and shrubs to annual grasses less resistant to drought stress. Deep-rooted leguminous trees and shrubs have also been increasingly harvested and burnt for fuel, and their role in water and nutrient cycling has diminished. Other species that depend on them for shade and nutrients cannot survive. The simplified soil and root structure is less able to absorb the moisture of seasonal storms, and the subsequent rapid runoff accelerates soil erosion, further inhibiting recovery.

Soil compaction and crusting, loss of soil organic matter, reduced soil-organism activity, and nutrient deficiency and imbalance reinforce one another in a cycle of resource deterioration (Lal, 1988). The interrelated effects of these conditions can be subtle. Soil erosion, for example, removes niches in which seeds germinate. Reduced numbers of trees and shrubs mean not only fewer seeds, but fewer birds and insects to spread seeds and pollen. Moreover, many trees must have their seeds pass through goats or camels before they can germinate. By such circuitous routes can the erosion of soil by wind and water, and the attendant loss of biological diversity, lead to land degradation and desertification throughout the world's arid regions.

In hill lands, the pressure of increasing population and the demand for land and fuel also lead to resource degradation, more marked because sloping land accentuates runoff and erosion (Jodha, 1990). Extensive deforestation can also affect entire watersheds. Reduced moisture retention in their upper basins can cause changes in the annual flood regimes of mighty rivers, such as the Nile, including severe flooding, and greatly reduced flow when water is most needed.

In input-intensive systems, such as the irrigated rice and wheat systems of Southeast Asia, high-yielding varieties produce two or more crops a year, with generous applications of fertilizers and pesticides. Recent reports (Byerlee, 1990; Ruttan, 1989) have described problems associated with maintaining current production levels, including the mining of trace nutrients, declining incremental response to increased fertilizer use, pest resistance, and reduced returns from additional research investment. In many input-intensive systems, water quality and availability are critical issues. In inadequately drained areas, irrigation is leading to salinization and consequent loss of productivity; in other areas, aquifers are being depleted. Contamination of groundwater is not yet as important a factor in developing countries as it is in some industrialized countries, but fertilizer and pesticide contamination of irrigation and other surface waters is important where these waters are also sources of drinking water or used for fish production.

The interrelated issues of population growth, intensified land use, environmental decline, and agricultural productivity at local and regional levels raise concerns about food security and quality, public health, and other long-term development problems. The issues are pertinent in all regions, but they are of special concern in the developing nations of the tropics, where the economic constraints and the development needs of rapidly growing human populations are most pressing. There, as elsewhere, environmental quality and development can no longer be separately considered. A quality environment and a healthy, stable resource base are essential for economic development, especially agricultural development. Conversely, ensuring a quality environment and resource base depends on changes in development policy and agricultural practices.

CHARACTERISTICS OF SUSTAINABLE AGRICULTURE AND NATURAL RESOURCE MANAGEMENT SYSTEMS

The concept of sustainable agriculture is a relatively recent response to interrelated environmental and economic concerns. Early discussions stressed the importance of maintaining the renewal capacity of agricultural ecosystems and claimed that many conventional agricultural practices were detrimental to that capacity. From further discussion has emerged an approach to agriculture that incorporates the principles of ecology by emphasizing interactions among and within all the components of agroecosystems.

As more individuals and organizations have begun to recognize the need for adjustments to conventional agriculture to make it environmentally, socially, and economically viable, sustainable agriculture has come to connote approaches to agriculture that can provide for the needs of current and future generations while conserving natural resources. Indeed, a major development in the past decade has been the emerging recognition on the part of agricultural production and environmental management groups that they share common, rather than competing, goals. In this context, sustainable agriculture is often used to refer to agriculture and all its interactions with society and the greater environment; as such, sustainable agriculture can be considered a vital component of current discussions of sustainable development.

The definition of agricultural sustainability, it is frequently noted, varies by individual, discipline, profession, and area of concern. The literature offers hundreds of definitions of sustainable agriculture. Virtually all definitions, however, incorporate the following characteristics: long-term maintenance of natural resources and agricultural productivity, minimal environmental impacts, adequate economic returns to farmers, optimal production with minimized chemical inputs, satisfaction of human needs for food and income, and provision for the social needs of farm families and communities. All definitions, in other words, explicitly promote environmental, economic, and social goals in their efforts to clarify and interpret the meaning of sustainability. In addition, all definitions implicitly suggest the need to ensure flexibility within the agroecosystem in order to respond effectively to stresses.

The characteristics of sustainable agriculture provide a framework and suggest an agenda for the perpetual dynamic evolution of agriculture to meet the needs of changing societies and environments. Sustainable agricultural systems must maintain and enhance biological and economic productivity. The former is required to feed individual farm families and the nonfarm population. The latter is required to provide income for farmers and low-cost food for consumers. Ruttan (1988) has pointed out that, for both the developed and developing world, "any definition of sustainability . . . must recognize the need for enhancement of productivity to meet the increased demands created by growing populations and rising incomes." Others emphasize that enhanced productivity cannot be gained at the expense of the resource base, but in fact depends on constant conservation efforts. "High rates of soil loss are causing declines in soil productivity worldwide, and most nations do not have sound land use policies to protect their soil and water resources. . . . The limited availability of fossil energy resources and their cost, which is expected to increase, make it unlikely that fertilizers and other inputs can offset severe land and water degradation problems, especially in impoverished nations" (Pimentel et al., 1987). Especially as the availability of new arable lands decreases, sustainability will require continual enhancement and improved management of soil and water resources and the protection of biodiversity in the system.

Sustainable agricultural systems should be both stable and resilient. Stability reduces risk and leads to continuity in income and food supply by fulfilling the short-term needs of farmers without incurring long-term environmental costs. Resilience permits adaptation to changes in the physical, biological, and socioeconomic environments. Sustainable agricultural systems should be environmentally acceptable; they should avoid erosion, pollution, and contamination, minimize adverse impacts on adjacent and downstream environments, and reduce the threats to biodiversity. Sustainable agricultural systems should also be economically viable in both the short and long term. Finally, they should be socially compatible with local people and political economies.

THE RESEARCH CHALLENGE:
ADOPTING A SYSTEMS-BASED APPROACH

Fundamentally, achieving sustainable agriculture under the mounting pressure of human population growth will demand that the world's agricultural productive capacity be enhanced while its resource base is conserved. If the well-being of the world's less advantaged people is to improve in any lasting sense, long-range concerns about food security and the health of natural resources must be addressed in planning future economic and social development. Research will be essential to this task. More specifically, researchers must devote greater attention to developing integrated cropping, livestock, and other production systems—and the specific farming practices within these systems—that enhance (or, at minimum, do not degrade) the structure and functioning of the broader agroecosystem.

A primary objective of research on sustainable agriculture and natural resource management is the integration of information in its application to the problems of agricultural development (Edwards, 1989; Edwards et al., 1990; Grove et al., 1990). This process requires an approach to interdisciplinary research that includes the following: (a) identification of the components and interactions that determine the structure and functioning of the agroecosystem as a whole; (b) formulation of hypotheses that focus on those components and interactions within the entire agroecosystem; (c) examination, testing, and measurement of the hypotheses; and (d) interpretation of results as they pertain to the various components of the agroecosystem and to the system as a whole. A lack of understanding of the interrelatedness of system components has undermined agricultural sustainability in the past, and failure to consider any one of them fully will inevitably undermine it in the future. A systems approach to research is necessary if these shortcomings are to be overcome.

In the United States, the lack of systems research has been identified as a key obstacle to the adoption of alternative farming practices and as a necessary step in the development of a more sustainable agriculture (National Research Council, 1989a, 1989b). In the even broader realm of international sustainable agriculture and natural resource management, the integrated research design, interdisciplinary participation, and systemwide perspective that the systems approach entails are necessary if the complex nature of sustainability is to be comprehended, the scientific basis of sustainability understood, and the threats to sustainability identified and addressed (Edwards, 1987).

Although the value of systems approaches has been increasingly recognized over the past decade, few crop and livestock production systems have been studied in detail. Agroecosystems are extremely diverse and variable, and thus the identification phase of research—the description of major components of the particular agroecosystem and the regional factors that act as constraints—is crucial.

A simple conceptual framework for the conduct of integrated agricultural systems research includes the following elements:

• description of the target agroecosystem, including its goals, boundaries and components, functions, interactions among its components, and interactions across its boundaries;
• detailed analysis of the agroecosystem to determine constraints on, and factors that can contribute to, the attainment of social, economic, and environmental goals;
• identification of interventions and actions to overcome the constraints;
• on-farm experimentation with interventions; and
• evaluation of the effectiveness of newly designed systems, and redesign as necessary.

Techniques for describing agroecosystems have been reported in the literature (for example, Clay, 1988; Conway, 1985). A description of the agroecosystem components and boundaries is essential in providing a focus for study, but it should not limit understanding of interactions with adjacent ecosystems, or with local, regional, national, and international political economies. The description of the target agroecosystem must be based on discussions with farmers and other local sources of information and the recommendations of scientists from the range of relevant disciplines. Description of the components of an agroecosystem is the traditional occupation of many agricultural scientists, but description and analysis of interactions among its components require farmer participation as well as an interdisciplinary perspective and a whole-systems approach. Because proposed interventions are aimed at assisting farmers in attaining their goals, understanding these goals is especially important.

Although the descriptive phase of sustainable agricultural systems research is largely qualitative, the analytic stage takes maximal advantage of quantitative information. The proposed descriptions may lead to hypotheses that require experimental study for resolution and quantification. For example, if nitrogen is suspected to be a limiting factor, then nutrient-response studies may be required. If losses to pests are hypothesized as an important factor, they can be quantified experimentally, and integrated management measures can be recommended for the pests identified. The result of the analytic phase is a more precise understanding of the factors that affect the attainment of the farmer's goals.

The design phase involves forming hypotheses about appropriate interventions that can contribute to the realization of the farmers' goals. It is a deductive process based on the description and analysis of the system. The final design represents the best collective judgments of the researchers and the participating farmers.

The evaluation phase assesses the interventions empirically and leads to further modifications. Effects must be measured in terms of the goals of the system, and trade-offs among goals must be determined for any proposed intervention. Interdisciplinary involvement and participation are essential in a successful evaluation phase.

As descriptive and analytic processes are employed in the study of agroecosystems in different regions and agroecological zones, the commonalities among them need to be emphasized and examined to elucidate their role in the functioning of the systems. Biological diversity, for example, is important to topsoil retention, nutrient cycling, and pest management in all agroecosystems. As these commonalities become better understood, they are likely to lead to global principles for the design of sustainable agricultural systems. The influence and importance of the commonalities may vary among agroecosystems, but research on them should be a high priority in all agroecosystems. Interdisciplinarity and integration will be fundamental to this effort.

2

Expanding the
Management Challenge

The urgent need for research on international sustainable agriculture and natural resource management, and for integrated approaches to that research, led Congress to direct the Agency for International Development (AID) to establish a new collaborative research support program (CRSP) to help lay the foundation for developing sustainable agricultural systems. This decision parallels recent developments within the international agricultural research center system and other agricultural research institutions (Consultative Group on International Agricultural Research, 1989, 1990). Forestry, sustainable agriculture, and other areas of natural resource management are gaining greater recognition within these institutions and a more prominent place on their research agendas.

The new CRSP would become the centerpiece of a comprehensive research program on Sustainable Agriculture and Natural Resource Management (SANREM) at AID that would involve U.S. and other developing country university researchers. It would also offer new opportunities for university researchers to work on these issues with colleagues from existing CRSPs, the international centers, national agricultural research systems, and private voluntary, nongovernmental, and commercial organizations.

HISTORY AND EVOLUTION OF THE
COLLABORATIVE RESEARCH SUPPORT PROGRAMS

During the late 1960s and early 1970s, serious concern arose regarding population growth and the demands that growth would place on the food production capacity of all developing countries. Discussions about "impending food crises" gained media attention. Meanwhile, a grass-roots effort in the U.S. land-grant universities grew, centering on the question of how they could most effectively assist developing countries in resolving food availability problems. The universities, with their rich experience in agricultural research, had proved their ability to improve the productivity, distribution, and utilization of land and water resources and were anxious to share their expertise.

Building on this groundswell of interest within the university community, AID identified a new, long-term mechanism for involving the land-grant universities in international agricultural research. In 1975, Congress passed the International Development and Food Assistance Act, Title XII of which authorized the president to "provide assistance on such terms and conditions as he shall determine . . . to provide program support for long-term collaborative university research on food production and distribution, storage, marketing and consumption." The act also provided that "programs under this title shall be carried out so as to . . . take into account the value to United States agriculture of such programs, integrating to the extent practical the programs and financing authorized under this title with those supported by other Federal or State resources so as to maximize the contributions to the development of agriculture in the United States and in agriculturally developing nations." This was the legislative foundation of the eventual CRSP structure (Yohe et al., 1990).

Between 1977 and 1982, the Joint Research Committee of the Board for International Food and Agriculture Development, which advises AID on university involvement in cooperative research, helped AID design and implement the eight existing CRSPs. The Joint Research Committee, which comprises AID and Title XII university representatives, was made responsible for oversight of Title XII research programs. The effort was unprecedented. Each time it approved a grant for another CRSP, the committee operated on the cutting edge of new experience. It allowed flexibility in the planning of each program, recognizing that initiatives addressing diverse concerns could not effectively be designed according to a standard pattern.

The CRSPs have since evolved into research enterprises involving U.S. universities, AID and its regional bureaus and overseas missions, other U.S. federal agencies, national agricultural research systems in developing countries, international agricultural research centers, private agencies and industries, and developing country institutions (Yohe et al., 1990). The eight CRSPs are conducting research on: (1) fisheries stock assessment, (2) human nutrition, (3) beans and cowpeas, (4) peanuts, (5) pond dynamics and aquaculture, (6) small ruminants, (7) sorghum and millet, and (8) tropical soil management. These programs involve more than 700 experienced international scientists from 32 U.S. universities and 80 international research institutions.

The design of the CRSPs reflected the understanding that international collaboration was key to successful agricultural research. The structure and organization of the CRSP model exemplify this internationalization of agricultural research. The host country and U.S. researchers share in the identification of research needs, the design of experiments, and the analysis of results. Collaborative research is jointly planned, implemented, and evaluated. The concept of networking is used to involve people and organizations not formally tied to a CRSP. The CRSPs use these networks to provide training through degree and nondegree programs and to establish long-term researcher-to-researcher links. Shared resources, peer review, and institutional support are critical to the success of their efforts.

The CRSP scientists carry out agricultural research and training activities that focus on identified constraints to food production, storage, marketing, and consumption. Their research and training address agricultural policy and planning, natural resource management, plant and animal improvement (including basic genetics, applied genetics, and biotechnology), plant and animal physiology and improved production practices, plant and animal protection, socioeconomic and cultural factors influencing production and consumption patterns, cultural constraints to technology adoption and development, improved food processing and household food security, and human nutrition. These programs place particular emphasis on the needs of small-scale producers and the rural and urban poor.

The CRSP concept has evolved into the effective mechanism its designers intended it to be and is producing significant benefits for both U.S. and developing country agriculture. The CRSPs have established long-term professional relationships that promote human resource development. In a relatively short time, these research programs have transcended political change, economic upheaval, environmental disasters, and institutional weaknesses to become one of the primary vehicles for U.S. involvement in international agricultural research.

CRSP INVOLVEMENT IN SUSTAINABLE AGRICULTURE

Sustainable agriculture is an evolving concept, and the furthering of the concept itself is a critical part of the overall mission of the proposed SANREM program. Since their inception, however, the CRSPs have implicitly addressed aspects of sustainability. Areas of research in which they contribute directly to sustainable agriculture include soil and water management, cropping systems, sustainable small ruminant production systems, aquatic production systems, coastal marine production and conservation, biodiversity protection and germplasm conservation, crop utilization systems, integrated pest management, and household food security. The experience of the CRSPs in these areas will undoubtedly continue to yield important fundamental lessons and knowledge.

Achieving sustainability in the developing world will always depend on the availability of a strong scientific and technical human resource base from which sustainability issues can be addressed. This is one of the chief contributions of the CRSP experience. The CRSP model, as noted, has promoted the long-term training and collaborative research relationships that help to build such a human resource base, to improve developing country research institutions, and to cultivate the integrated approach so necessary to work on sustainability issues. The components of this institutional development include human resource training and updating, operational research support, cohesive and continuous commitment, long-term networking with peer scientists, multi-institution research integration, interdisciplinary research integration, and inter-CRSP research integration and collaboration.

Sustainability and agroecological considerations are so important and central to attaining development goals that they should be fundamental to planning and carrying out all the agricultural and natural resource programs that AID supports. The new CRSP, then, should not be viewed as the only AID sustainable agriculture activity; all other AID-supported activities, including the existing CRSPs, address various aspects of sustainability, and they must continue to do so. The new CRSP should complement these existing efforts and add a critical dimension of integration as the core activity of a comprehensive SANREM program.

3

Considerations and Criteria for the SANREM Program Design

The establishment of the Sustainable Agriculture and Natural Resource Management (SANREM) program would foster creative approaches to sustainable agriculture and natural resource management at the Agency for International Development (AID). It should build on the efforts of previous programs and work as far as possible with them in defining sustainability issues that involve as much of AID's research and related activities as possible. The nature of this task would require the SANREM program to adopt an approach to research that integrates the various disciplines in determining priorities that focus on the health of the entire agroecological system.

A research proposal and granting structure for the SANREM program should encourage the above qualities by providing a framework for the optimal mixture of specialized expertise and systemswide perspective, as outlined previously. In the review of grant proposals, weight should be given to creativity in the design of research that promises new insights into the physical, biological, and chemical bases of agroecosystem interactions; that examines the impact of those interactions on productivity; and that addresses social, cultural, environmental, and institutional issues in innovative ways.

The research approach must take into account those factors that influence the ability of people to improve their livelihood, income, and health. It must make use of and strengthen existing pools of indigenous knowledge available for the design and adoption of sustainable production systems. Research projects should seek to understand how physical, biological, economic, and social factors interact and must be balanced to manage agroecosystems in a sustainable manner. The SANREM program should primarily seek to promote research that adds to this understanding and that works with the farmer and across disciplines and institutions to fashion the tools, perfect the techniques, and design the farming systems that can shape a sustainable future.

Suggestions for research in four agroecological zones (the humid tropics, semiarid range and savannah, hill lands, and input-intensive systems) are summarized in Appendix D of this report. This material is included with the caution that, in focusing attention on a specific agroecosystem, the broad commonalities among all agroecosystems and their interrelationships must be kept in mind. In SANREM program activities, the agroecological zone should serve mainly as a tool for organizing and implementing new strategies in the investigation of common properties and processes; namely, the functions of on-farm and off-farm biodiversity; soil and water management; the role of biological nutrient flow and cycling in enhancing fertility; and the human dimension of sustainability, including especially the role and impact of farmer-consumer relationships, infrastructure, institutions and their management, land tenure patterns, gender roles, and agricultural and natural resource policies and programs. Similarly, in all agroecosystems, inputs can be characterized according to their nature and impact. Within each zone, the level (high or low), source (farm or purchased), and relationships of inputs should be addressed in the experimental design.

PROGRAM OBJECTIVES

The primary aim of the proposed SANREM program is to stimulate and support innovative, integrated systems-based research that will lead to the identification and development of sustainable agricultural production systems. This research must address all agroecological factors in devising cropping, livestock, and other food production systems—and specific farming practices within such systems—that are capable of improving human welfare, countering the detrimental effects of current agricultural practices and policies, and conserving natural resources as pressures on the global resource base increase. This effort will benefit not only the developing countries in which it is conducted and to which it is directed, but also the United States, through the development of more effective research methodologies, the training of U.S. researchers, and the acquisition of results pertinent to the sustainability of U.S. agriculture and natural resources.

The sustainability of any agroecosystem is influenced by many factors—social, economic, biological, and environmental. Some of the factors, such as nutrient management, the control of pests, and the influence of policies and institutions, are common around the world. Others are regional and require that questions be resolved and measures adopted on the basis of the unique geographical, ecological, historical, political, social, and economic circumstances at a given site. Sustainability implies the securing of a durable, favorable balance of economic and environmental costs and benefits within the context of the system as a whole. The objective is to increase the per capita productivity of farming systems and the long-term ability of the farmer to meet family, local, and regional livelihood and economic goals. Progress will ultimately depend on the ability to engage human ingenuity in the maintenance and enhancement of the natural resource base—its diversity, fertility, stability, and renewal capacity.

An integrated systems approach, whether formally or informally defined as such, will be essential to all research under this program. The research location should encompass a landscape or political unit of sufficient size and diversity to support studies of all the principal determinants of sustainability within the agroecosystem. To the fullest extent possible, farmers should actively participate in each phase of the research process, from initial planning and testing to technology development, dissemination, and other extension-related activities. Because considerable attention is already being given to input-intensive agroecosystems, efforts should be directed primarily, but not exclusively, to the more fragile agroecosystems.

A major aim of the SANREM program would be to design and field test systems of sustainable agriculture and natural resource management. Research, thus, must illuminate the principles and theory that underlie sustainability. Those general principles and theory can then be applied to specific situations across broad ecological zones. Knowledge of the effects of system structure is crucial to managing systems for biological stability, environmental protection, improved efficiency of resource use, and greater productivity. Research should test improved technologies for cropping systems. The knowledge needed can only be gained over a relatively long period of time—at least 10 to 15 years under most circumstances. However, specific test results and recommendations should be available within the first 3 to 5 years. At the same time, the problem focus will be sharpened, and crucial experience will be gained in assembling and managing complex international, multidisciplinary research efforts.

In carrying out these functions, the SANREM program will lend needed support and recognition to interdisciplinary research and the publication of results in peer-reviewed journals. The goal of sustainability and the scientific problems it raises are complex. Accordingly, research should involve natural, agricultural, and social scientists who have a commitment to interdisciplinary inquiry. This commitment must be shared by collaborating institutions and local governments if the program is to succeed.

The implicit involvement of students and other training activities should contribute to strengthening institutional capacities in the host country. It is expected that the SANREM program would include elements that have a significant degree and nondegree training component. The U.S. institution or consortium of institutions participating in the SANREM Collaborative Research Support Program should have the necessary scientific capability, field experience, and training capacity to form working agreements with relevant international and national institutions to effect the needed research.

CRITICAL AREAS OF INQUIRY

Research should take into consideration all the basic elements involved in agricultural system performance (including soil and water resources, tillage and cultivation methods, cropping patterns, animal husbandry, nutrient management, and pest management), but it should devote attention to additional components (such as aquaculture and farm forestry) as appropriate. Resource policies and other institutional factors play a critical role in determining the choices that farmers make and, hence, the sustainability of farming systems. Accordingly, research must also be directed to the socioeconomic and policy context within which farmers make their decisions.

Knowledge of all relevant components and their interactions is fundamental to understanding the functioning and management of agroecosystems. However, this knowledge is often inadequately integrated or lacking altogether. Greater understanding of the sustainability of agroecosystems will require that all relevant factors be researched, and that they be researched together. The approach to research therefore should emphasize the following cross-cutting ecological and socioeconomic concerns.

Four research areas are common to all agroecosystems, and they provide the framework within which projects can address the broad range of issues relating to sustainable agriculture and natural resource management. They are integrated pest management, integrated nutrient management, the social, political, and institutional context, and integrated institutional management.

Increased concern for environmental and human safety and for the long-term sustainability of agricultural production systems has given added incentive and importance to one area of research with a strong legacy of innovation—*integrated pest management* (IPM). Over the past 30 years, IPM has built a solid record of research and demonstration of pest management methods that are less costly and more flexible, reduce the human health and environmental effects of synthetic pesticides, successfully combat pest resistance, and help to ensure viable, consistent yields.

As the SANREM program seeks to advance sustainable agriculture and resource management, IPM will assume an even more critical role. Many of the regions where sustainability is most at risk are areas where pest pressures (from weeds, insects, and pathogens, as well as pre- and postharvest vertebrate pests) are most persistent and safe, affordable, and accessible control methods are most needed.

Traditionally, the aim of most IPM programs has been to use multiple chemical, biological, and cultural tactics to maintain pest damage below the economic injury level while providing protection against hazards to humans, animals, plants, and the overall environment. In practice, there has been a lack of true integration in managing inputs for the control of injurious arthropods, diseases, weeds, and other pests. To achieve this goal IPM must be integrated with sustainable agriculture and resource management. More research is needed into fundamental ecological relations and management techniques involving pests and their hosts, parasites, predators, and antagonists; cultural and biological pest controls; and other factors that determine the ultimate impact of pests.

Integrated nutrient management is concerned with the integration of chemical, biological, and cultural sources of nutrients essential for crop production. Although the concept is applicable in all systems, it is of particular importance, in an operational sense, to the poorer soils that predominate in the tropics. Traditional agricultural systems depend on the use of organic nutrient sources, including animal and green manures, crop residues, legume crops, crop rotations, agroforestry, and fallows. Such cultural methods provide other benefits, including improved soil tilth and water-holding capacity, enrichment of soil biota, more efficient binding and release of mineral nutrients, and protection against persistent weeds, diseases, and other pests. Dependence on excessive chemical inputs can have a negative effect on these important factors.

Much of the recent interest in sustainable agriculture has grown out of concern over the agronomic, environmental, and economic costs of increased reliance on off-farm sources of nutrient inputs. The authors of the 1989 National Research Council report *Alternative Agriculture* point out that "efforts to provide adequate nutrition to crops continue to be hindered by inadequate understanding and forecasting of factors that influence nutrient storage, cycling, accessibility, uptake and use by crops during the growing season. As a result, farmers often follow broad guidelines that lead to insufficient or excessive fertilization" (National Research Council, 1989a:144).

This situation is not unique to high-input cropping systems in the United States. Inadequate understanding of the ecological dynamics of nutrient cycling in all agroecosystems hinders progress toward more efficient and effective integrated nutrient management strategies. This progress must be achieved to take full advantage of all nutrient inputs—chemical, biological, and cultural—and to cut nutrient loss. Integrated nutrient and integrated pest management are basic to crop and animal integration for sustainability, and they relate directly to the important roles that biological diversity and the availability of organic matter play in sustainability.

The *social, political, and institutional contexts* within which on-farm and off-farm activities take place must be given full attention by researchers if they are to identify and suggest remedial steps that can help remove constraints to sustainability. This must include attention to land tenure issues, property rights, the social and environmental impacts of policy, and economic incentives and disincentives.

Attention to these concerns will demand a strong and innovative social science component in the research design, the focus of which should be the institutional and policy conditions that influence on-farm resource management patterns. This research should address issues of gender and age, the impact of production alternatives on social structure, and ways to strengthen critical human resources, including especially the base of native and indigenous knowledge. If the adoption of more sustainable methods and technologies should involve hardship for some local farmers, such results should be anticipated, forthrightly acknowledged, and studied with a view toward amelioration.

All of the considerations above suggest the need for *integrated institutional management,* including a production economics component. Such management is needed to guide the complex interactions between food and fiber production and the policy, trade, and political environment.

The four focus areas of SANREM research must proceed concurrently as research projects mature. Projects should focus attention on agroecosystems in a manner that enhances stability, environmental protection, and resource conservation. Work on integrated pest management and integrated nutrient management will be central to this effort in that they seek to understand technically how to optimize the use of on-farm and, where necessary, purchased inputs while conserving the soil and water resource base. The social science work will be central to understanding how the people in both farm and nonfarm sectors view the systems and to identifying the policies and incentives needed to sustain them. These perspectives must be integrated at the time research is initiated, and not added as an afterthought, if results are to be meaningful and applicable.

Research needs in these areas will depend on the specific site conditions and the specific changes required there. However, a broadened systems approach is needed to define specific needs and to apply the findings. The goal is to realize the biological production potential of the area while ensuring social and economic viability, environmental quality, and resource conservation. The trade-offs among environmentally friendly technology, enhanced farm family income, and increased capital or input investment can then be better understood. That understanding, in turn, will permit local and national decisions to be made according to development objectives.

4

SANREM Program Management and Grant Administration

A thoughtful and clearly articulated research agenda is crucial to the success of the proposed Sustainable Agriculture and Natural Resource Management (SANREM) program. No less important are an organizational structure that fits the components into a logical framework and a management device that promotes a sense of esprit de corps among program participants, while ensuring that research program responsibilities are met, administrative actions are orderly, and reporting is timely. Management of the SANREM program would also entail overseeing the international collaborative arrangements so essential to its success. In meeting these administrative requirements the program should adopt the essential features of, and be patterned closely after, the existing collaborative research support programs (CRSPs).

The CRSPs have been operational for more than 15 years. To meet changing priorities and funding constraints, the details of their internal structure and functional operations have been modified over time. Despite modifications, the same basic components for a collaborative program have been maintained. The CRSP model and experiences are valuable resources in designing the organizational framework and management approach for the proposed SANREM program.

ORGANIZATION AND MANAGEMENT OF THE SANREM CRSP

The SANREM CRSP should be organized along the lines set forth in *Guidelines for the Collaborative Research Support Programs* (Agency for International Development, 1985; the "guidelines" hereafter). However, some variations from the standard organizational framework are required.

The existing CRSPs focus on a commodity or discipline; the SANREM CRSP will be multidisciplinary. The existing CRSPs are funded through one administrative unit within the Agency for International Development (AID); the SANREM CRSP will be funded through multiple administrative units. The existing CRSPs have limited involvement by AID in programming; more substantive participation by the collaborating AID offices will be required in planning, programming, and implementing the SANREM CRSP if it is to be integrated with other programs of the Bureau for Science and Technology. Attention must be given to these unique aspects of the SANREM CRSP, and the guidelines provide the necessary flexibility to address them.

Administration by a Management Entity

The guidelines provide for the administration of each CRSP by a central agency, or management entity. They should be adhered to in administering the SANREM CRSP, with the following modifications.

Following the selection of the recipient for the research core grant (discussed below), representatives from each of the universities, institutions, and organizations involved in the SANREM CRSP will be asked to recommend interested candidate institutions, in rank order, to serve as the management entity. To be eligible, the management entity candidate must have the legal status of a juridical body. It may be a U.S. university, an administrative unit within a university, or a consortium or other structures of universities, legally organized as a juridical body representing the participating universities. An institution eligible to receive a federal grant would be eligible to serve as a management entity. The planning entity will use these recommendations in preparing its recommendation to AID for a management entity.

The management entity will be responsible for all aspects of CRSP management and will be the administrative link between the SANREM CRSP and AID. In discharging its responsibilities, the management entity will undertake the following duties:

• Receive and administer funds provided to support activities of the SANREM CRSP, including research support grants (see below).

• Enter into agreements with participating U.S. and developing country institutions to implement activities, and provide funding in accordance with the initial grants and subsequent modifications in activities and budgets.

• Establish a system for effective management that will ensure accountability in the use of funds for the intended purpose.

• Ensure that commitments for matching resources are met and accountable.

• Establish a system to facilitate and manage travel.
• Ensure that timely and effective reviews of CRSP activities are performed.
• Institute changes in activities and funding in collaboration with the board of directors, the technical committee (discussed below), and the AID program manager, as needed and appropriate.
• Conduct general oversight of the technical activities and provide leadership in initiating actions to consider modifications or additions.

The need to communicate in a useful form the substantive results emanating from the SANREM CRSP is an important responsibility. To address this need, the CRSP management entity should not only create and support a mechanism to provide the necessary periodic reports required by the funding agency, but also initiate other means of disseminating program results.

Supporting Units

To assist in carrying out its responsibilities, the management entity will form a board of directors, a technical committee, and an external evaluation panel. Establishment of a sustainable agriculture and natural resource management committee within AID is also recommended to coordinate the program planning and evaluation activities of the participating AID offices.

Board of Directors

The board of directors will consist of representatives from some or all of the participating institutions and may include individuals from other organizations. Members of the board should have some expertise in one or more of the disciplines involved in SANREM program research, some authority to represent the administration of participating institutions, and the ability to provide unbiased analysis of program strengths and weaknesses. The director of the management entity and the AID program manager will be ex officio members of the board. Responsibilities of the board include evaluating and recommending revisions in policies, programs, and budgets, and providing advice to the management entity on any matter that could improve functioning of the SANREM CRSP. The board will submit periodic reports of its findings and recommendations to the director of the management entity.

Technical Committee

The technical committee will be drawn primarily from the principal scientists engaged in the work of the CRSP. The director of the management entity and the AID program manager will be ex officio members of the committee. The committee's responsibilities will be to review technical matters pertaining to the SANREM program, to develop or review recommendations for modifications in the research program, and to develop or review recommendations for adjustments in funding. It will submit periodic reports of its findings and recommendations to the director of the management entity.

External Evaluation Panel

The external evaluation panel will consist of a minimum of three senior scientists recognized by their peers and selected, according to the procedures set forth in the guidelines, for expertise relevant to the SANREM program and experience in research or research administration. The responsibility of the panel will be to evaluate, as deemed necessary, the status, funding, progress, plans, and prospects of the SANREM program and make recommendations based on these evaluations in a report to AID.

Sustainable Agriculture Committee

The Agency for International Development is a major collaborator in program planning and evaluation for each CRSP. Because of the multiple AID offices and divisions involved with the SANREM CRSP, AID should establish a sustainable agriculture and natural resource management committee. The committee members would be drawn from the participating offices and selected to ensure representation for the biological, physical, and social sciences. One member of the committee would be the AID program manager for the SANREM CRSP and provide the primary link between the agency and the CRSP management entity. The committee would be responsible for reviewing and analyzing the program, recommending modifications and additions of activities and funding to AID and the CRSP management entity, and promoting links between the SANREM CRSP and other relevant activities of the Bureau of Science and Technology.

GRANT ADMINISTRATION

Grants should be awarded under the proposed SANREM program on a competitive basis to a limited number of institutions or consortiums.

Types of Grants

Three types of competitive grants should be made available under the proposed SANREM program: research planning grants, a research core grant, and research support grants.

Research Planning Grants

Applicants who intend to apply for the research core grant should be encouraged, but not required, to submit a preproposal to AID by July 1991 for a research planning grant. The purpose of the planning grants would be to support enhanced interdisciplinary interaction, on-site visits to potential study sites, and the development of links with cooperating institutions in the process of preparing and refining proposals for a research core grant. From the highest ranking preproposals submitted, a maximum of six planning grants of up to $50,000 each per institution or consortium should be awarded for proposal development during the initial year of the program. Recipients of planning grants would be required to submit to AID a full proposal for the research core grant. Alternatively, a report to AID describing the activities undertaken would meet this requirement. Planning grants are recommended because the type of integrated research necessary to fulfill the objectives of the SANREM program will require new modes of collaboration and are likely to involve institutions and individuals that may not have worked together before.

Research Core Grant

A research core grant should be awarded to support a long-term, fullscale interdisciplinary collaborative research program (the SANREM CRSP) on sustainable agriculture and natural resource management in one or more of the world's principal agroecosystems. This grant should be awarded in the second year of the program at a level of about $2.5 million per year (administrative expenses of $300,000 per year would be included in this amount).

The initial core grant should be authorized for 5 years. Prior to the end of the third year, however, a comprehensive review should be undertaken and a decision made to extend or terminate the grant. The review would be conducted according to the procedures in the CRSP guidelines, with such modifications as agreed to by AID and the management entity. Funding schedules should be in accordance with AID administrative actions. The core grant recipient would be required to match with nonfederal resources (cash or in-kind contributions) an amount equal to not less than 25 percent of the federal funds provided, except for those costs paid by federal funds that have been determined to be exempt from these requirements.

The core grant recipient should be selected from the pool of final proposals, which should be open to all qualified applicants. A recommended timetable for proposal submission and the awarding of grants is provided in Table 4-1.

Research Support Grants

Research support grants should be awarded to support research of direct

TABLE 4-1 Recommended Timetable for Awarding Grants Under the Proposed Sustainable Agriculture and Natural Resource Management (SANREM) Program

Program Phase	Target Date
YEAR 1	
National Research Council recommendations to the Agency for International Development (AID)	February 1991
Request for proposals for planning grants and type B research support grants (RSGs) distributed by AID	May 1991
Proposals submitted to AID for evaluation	July 1991
Proposal review and selection process completed	September 1991
AID approval of planning grants and type B RSGs	September 1991
YEAR 2	
Full proposals for core research grants submitted to AID	January 1992
Proposal review and recommendation of core research grantee	March 1992
AID approval and award of core grant (core grantee institutions select management entity)	April 1992
Type A and additional type B RSG proposals solicited	May 1992
RSG proposals submitted to AID and management entity	July 1992
RSGs awarded by AID and management entity	September 1992
YEAR 3	
Additional RSG proposals (types A and B) solicited by AID and management entity	January 1993
RSG proposals submitted	June 1993
RSGs awarded by AID and management entity	September 1993

NOTE: This timetable is for the first 3 years of a long-term program; after year 2, the timetable would be determined by the management entity and AID.

and immediate relevance to the goals of the SANREM program within other collaborative research programs, including existing CRSPs. This mechanism would permit the SANREM program to have access to research on aspects of sustainability within the current CRSPs and other AID-funded research projects. These grants would support research of value to the SANREM program and would integrate results from other AID-funded research.

Two types of research support grants are recommended: type A, to be awarded by the CRSP management entity, beginning as soon as the SANREM CRSP is established; and type B, to be awarded directly by AID's Bureau for Science and Technology, beginning as soon as possible. A limited number of these grants of up to $100,000 a year for an initial 3-year period should be awarded. The awarding of type B grants should neither hinder nor promote the eligibility of the same institution for the core grant.

To achieve their purposes, the research support grants should be administered partly by the CRSP to support close integration of institutions or individuals with expertise of particular relevance to the core research (type A), and partly by AID itself (type B) to support integration of other AID-funded research activities with the SANREM program. Selection of recipients for the two types of research support grants should be made by AID, based, respectively, on two types of recommendations: from the management entity of the SANREM CRSP (type A), and from a special peer review panel constituted for this purpose (type B). The number of research support grants will depend on the level of funding available each year from the Bureau for Science and Technology and from AID's regional offices and missions and other sources (buy-ins). If the SANREM program is allocated the $10 million over 3 years recommended by Congress, after completion of the planning phase (year 1), about $2.5 million per year should support the CRSP through the core research grant, with another $300,000 per year for the type A research support grants. AID may award the remaining annual allocation (and buy-ins) directly to other grantees as type B research support grants in support of SANREM program objectives.

Funding Levels

The levels of funding recommended above are based on several considerations. For the initial phase of the SANREM CRSP, Congress recommended, and AID has used as the basis for planning, a level of $10 million over 3 years (an average of about $3.3 million a year). Congress and AID are to be commended for acting quickly to establish a new SANREM CRSP, but adequate support for research on sustainable agriculture and natural resources management over the long term will require considerably higher levels of funding. Funding must be sustained for at least 15 to 20 years to achieve program objectives. Funding from other AID offices (buy-ins) should be sought as a means of integrating SANREM research and other AID-funded research projects and programs, including the existing CRSPs. Close collaboration with researchers who are funded by other donors should be sought.

In this context, at least six research planning grants may be necessary to encourage the desired institutional participation and innovation during the initial stages of the program. Given that coordination, including foreign travel, will be required among several participating institutions, $50,000 (per institution or consortium) may be needed to provide for the travel and staff time that a quality proposal will entail.

Using research support grants to add an integrative dimension to existing collaborative research is a critical and innovative part of the proposed SANREM program. Since two types of research support grants are envisioned, to be administered by the CRSP management entity and directly by AID for their respective collaborating purposes, a minimum of eight support grants (at about $100,000 each per year) is recommended for the initial 3-year period.

If $800,000 is allocated annually for research support grants, after year 1 an average of $2.5 million per year should be available for the CRSP. Although this amount could be divided among more than one core grantee, the broad range of issues that must be covered and the number of institutions that might be involved in the type of interdisciplinary research required suggest that it would be preferable to fund, on a competitive basis, a single institution or consortium with $2.5 million—$2.2 million for research and $300,000 for administrative expenses of the management entity.

Institutional Participation

Research conducted under the SANREM program would necessarily demand a broad range of expertise and international experience in the natural, agricultural, and social sciences. To be successful, projects may require the involvement of organizations and institutions that are not currently participating in Title XII programs. The program should be structured to allow both Title XII and non-Title XII universities to receive program funds, and to encourage the participation of other groups with relevant experience and expertise, including private voluntary, nongovernmental, and other private sector organizations. This goal could be achieved through contractual arrangements between universities and nonuniversity collaborators. The SANREM program should capitalize on the research and development capabilities of the entire U.S. system and of diverse collaborators in developing countries. Innovative collaborative arrangements, especially with relevant host country entities, should be encouraged to meet this goal.

CONTENT OF RESEARCH PROPOSALS

To meet essential criteria for funding, research proposals submitted for funding under the SANREM program should be required to provide information and demonstrate capacities as indicated below. Proposals for research planning grants and the research core grant should meet the same set of requirements to the fullest degree possible. Research support grant proposals, on the other hand, should meet those requirements from among the following topics that are necessary to augment their established research agenda.

Description of Research Location and Site

Proposals should describe the specific region, area, and agroecosystem in which the research will be conducted. This section should address the field research location, its facilities, the availability of local expertise, other support available from the collaborating country, and the potential for local outreach activities.

Significance of Research and Site

Proposals should explain the local, regional, and global significance of the study area, the type of cropping system(s) to be studied, and the socioeconomic and biological interactions chosen for investigation and analysis.

Problem Description and Research Methodology

Proposals should describe specifically the biological, ecological, social, and economic aspects of sustainability, and the constraints to sustainability, that the proposed research will help to elucidate.

Proposals should define specifically the on-farm, landscape, or regional problems to be addressed, the hypotheses to be tested, and the experimental approach that will be used to test the hypotheses and to identify, investigate, and ultimately overcome the constraints to sustainability. Proposals should also describe how the proposed research will fill gaps in existing knowledge and ongoing research. They should further define how hypotheses and research results will be integrated into an evolving theory of research design.

Proposals should also give evidence of attention to the special concerns described in AID's Program in Science and Technology Cooperation proposal guidelines (Agency for International Development, 1990). All AID-funded research requires a description of the steps the investigators will take to eliminate hazards or conform to ethical or environmental practices that are prescribed by law or scientific convention. These requirements include the following:

• A protocol and informed consent form for any research involving human subjects or volunteers. (Appropriate certification of ethical review committees may also be required.)

• Protocols describing the safe handling and disposal of any material presenting a hazard to research personnel, including radioisotopes, toxic chemicals, mutagens, or human pathogens. Proper containment and disposal procedures for any plant or animal pathogens are also required, along with certification of institutional approval for the protocol.

• Notification procedures for any international shipments of biological material, including submission of copies of the appropriate import and export permits.

• Description of how recombinant DNA work undertaken in the project will adhere to established guidelines (for example, those of the National

Institutes of Health) and have certification of institutional approval. Prior to field release of any engineered recombinant organisms, a detailed protocol in conformance with established U.S. and collaborating country guidelines must be approved by an institutional review committee.

• Description of how research involving animals will follow established guidelines that ensure their humane treatment. If any endangered or potentially endangered species (animal or plant) are to be used, a protocol describing efforts to mitigate the impact on this species must be provided, along with certification of governmental approval.

• A protocol for hazard reduction if the research will generate any environmental concerns, either at the research stage or with widespread application of the results. Such hazards may be physical, chemical (for example, pesticides), or biological in nature.

• Indication of the sharing of rights (including ownership and control) among the collaborators if the results of the research are likely to result in a product or process for which intellectual property rights are applicable. Preexisting intellectual property rights should also be considered. A process must be established, if not already in place, to protect property rights if a patentable product or process is developed.

Systems-Based Approaches to Ecological and Socioeconomic Research

Proposals should demonstrate a systems-based approach to research.

Proposals should place special emphasis on integrated pest management and integrated nutrient management. They should describe the relationship between ecological factors and policy and institutional factors within the selected agroecosystem. Accordingly, the research design should include innovative social science components that focus on these institutional and policy factors as they influence on-farm management decisions and patterns.

Proposals should also include a detailed plan for taking into account the cultural, economic, and indigenous knowledge characteristics of the region, both in the development of research and in prospective strategies for the refinement and exchange of technology among farmers. This plan should also indicate the relevance of the work to researchers and research institutions that are developing sustainable management systems within the area or in comparable areas elsewhere in the world.

Collaborative Arrangements Among U.S. and Host Country Institutions

Proposals should describe plans for facilitating collaboration among participating U.S. institutions, including institutional responsibilities, logistical arrangements, and communications. Plans should be indicated for the involvement of farmers, host-country scientists, members of private voluntary and nongovernmental organizations, and extension workers in the design, implementation, and dissemination of research results.

Information About Researchers and Other Collaborators

Proposals should include names and biographical information describing the qualifications of all researchers and other collaborators who will contribute substantially to the proposed research in the United States and host country.

Capacity for Interdisciplinary Research

Proposals should include a description of the team members' experience in, and specific plans for, the implementation and support of interdisciplinary research. At a minimum, proposals should address the following aspects of interdisciplinary research (Melcher and Stanbury, 1990).

Logistics

Proposals should provide detailed information about how the program will be set up to foster interdisciplinary collaboration, including composition of core staff, location of main facilities, composition of field teams, and location of research sites (farm or experiment station). They should also indicate proposed methods for transferring experimental methodologies to farmers' fields and how the methodologies will be adapted for and adopted by farmers. Finally, they should indicate awareness of the time demands of interdisciplinary research by specifying realistic time frames.

Team Building

Proposals should specify mechanisms for identifying and overcoming disciplinary biases and assumptions. The type of team-building methods (such as diagnostic analysis or the team planning meeting approach) that will be used to promote collaboration and interdisciplinary interaction in all relevant phases of the research, including fieldwork, should be indicated. Proposals should demonstrate prior experience in building interdisciplinary teams, including the use of facilitators and other professionals who can expedite the process of team building from the outset.

Objectives, Goals, and Performance Indicators

Proposals should indicate a set of objectives and goals common to both the social and biophysical sciences and a set of performance indicators by which to judge project "success" in terms common to all disciplines. Some of the most significant problems in interdisciplinary research arise when the performance indicators are not clear across disciplinary lines. In an effort to avoid this, proposals should specify a preliminary list of performance indicators that demonstrate a capacity for interdisciplinary research evaluation.

Data Collection, Management, and Information Sharing

Because interdisciplinary research has the potential to generate large quantities of data, proposals should indicate plans for data management, for information sharing among team members, and for formal publication. "Quick and clean" methods, such as rapid rural appraisal, or the development of a comparable methodology, are encouraged when appropriate. Proposals should also identify the characteristics and handling of a minimum data set on sustainable agriculture.

Feedback Mechanisms and Capacity for Flexibility

In addition to drawing on the theoretical and methodological strengths of individual disciplines, the research methods proposed should reflect a capacity for creative synthesis among disciplines and sufficient flexibility to accommodate the pursuit of multiple goals, prevent premature closure, and permit adaptive iterative changes as the research progresses. Proposals should indicate specific methods, such as a "learning process approach," for responding to and learning from other scientists and farmers and, where relevant, adjusting research priorities.

Input from Sustainable Agriculture Professionals

In addition to interactions among research team members, proposals should indicate how the team members will draw on local expertise as well as the broader research community. They should include such activities as workshops for discussions with representatives of a broad range of disciplines and organizations (including private voluntary and nongovernmental organizations, developing country experts, and private research centers).

Farmer Participation

Proposals should specify procedures for soliciting and ensuring farmer participation in all phases of the research, from problem identification and establishment of research priorities to evaluation and dissemination of results. They should specify mechanisms for ensuring that farmer participation is sustained, and they should indicate some possible ways of communicating with farmers more effectively (such as by developing local farmer organizations or by identifying progressive or experimental farmers who can disseminate new sustainable agricultural methodologies).

Institutional Sustainability

Proposals should specifically address the problem of sustaining interdisciplinary research among the U.S. and host country research staffs and within host country institutions. The proposal should demonstrate that the proposed methods will not be "one shot" interdisciplinary efforts at host country sites, but rather will foster new interdisciplinary approaches and promote policy changes within the national institutions themselves. Applied research on the bureaucratic agencies and host country research institutions involved in promoting sustainable agriculture—the management structure, system of incentives, reporting lines and communication between offices, and interdisciplinary training for host country staff—should be high-priority areas. Similar issues regarding the U.S. institutions are also of interest.

Broader Issues and Impacts

Proposals are also encouraged to indicate how they will set their research in a broader context—that is, to specify how the proposed research will be linked with its broader social, political, economic, and environmental contexts. They should also indicate how they will assess technology needs, the expected consequences and the impacts of the proposed research, and the distribution of associated costs and benefits.

Capacity to Develop Technologies and Disseminate Knowledge

Proposals should show how evolving systems theory and research findings will be translated into situation-specific research design and on-farm practices. They should also describe past accomplishments of the principal investigators and their institutions, and they should demonstrate special expertise in systems-based research related to the physical, biological, ecological, social, and economic bases of sustainable agriculture and natural resource management.

Budget

Proposals should include budget details, indicating not only the amount of funding requested, but also the contributions from the various collaborating institutions and other matching funds. As in the existing CRSP partnerships between AID and U.S. universities, a minimum match of funds of 25 percent through cash or in-kind contributions is required from U.S. institutions.

CONCLUSION

The establishment of the proposed SANREM program, and the competitive grants it would make available, would provide critical support for collaborative research on sustainability in developing countries. In particular, the SANREM program is designed to encourage imaginative approaches to interdisciplinary research on sustainable agriculture and natural resource management. It is expected that the SANREM program will attract a wide range of U.S. and foreign talent. Although the need for new approaches, innovative experimental design, and integrated training has long been recognized, the institutional and financial means to implement responses have been scarce. Research of the kind needed is long term and complex, requiring sustained commitment. Congress and AID are commended for their initial investments in the new CRSP and for their support of its goals. Although a modest step given the extent of the challenge, the creation of the SANREM program would demonstrate the effectiveness of new approaches and catalyze support from other parts of AID and from other donor agencies.

In the longer term, the SANREM program is expected to generate research results that can contribute directly to developing sustainable agricultural systems and natural resource management strategies. The understanding gained through SANREM research will advance both the theory that underlies sustainability and the design of practices that promote sustainability at the farm, landscape, and agroecosystem levels. In the process, fruitful research relationships will be created that promise to establish enduring international partnerships.

References

Agency for International Development (AID). 1985. Guidelines for the Collaborative Research Support Programs Under Title XII. Washington, D.C.: AID.

AID. 1990. Special concerns review list. Office of the Science Adviser, Agency for International Development, Washington, D.C. Photocopy.

Byerlee, D. 1990. Technical change, productivity, and sustainability in irrigated cropping: Emerging issues in the post-Green Revolution era. CIMMYT Working Paper 19-7. Mexico: Centro International de Mejoramiento de Maíz y Trigo.

Clay, J. W. 1988. Indigenous Peoples and Tropical Forests: Models of Land Use and Management from Latin America. Cambridge, Mass.: Cultural Survival.

Consultative Group on International Agricultural Research (CGIAR). 1989. Sustainable Agricultural Production: Implications for International Agricultural Research. Rome, Italy: Food and Agriculture Organization of the United Nations.

CGIAR. 1990. Sustainable agricultural production: Final report of the CGIAR committee. MT/90/18. Presented at the CGIAR Meeting, May 21–25, 1990, The Hague, Netherlands. Photocopy.

Conway, G. R. 1987. Properties of Agroecosystems. The Hague, Netherlands: Elsevier.

Crutzen, P. J., and M. O. Andrae. 1990. Biomass burning in the tropics: Impact on atmospheric chemistry and biogeochemical cycles. Science 250:1669–1678.

Edwards, C. A. 1987. The concept of integrated systems in lower input/sustainable agriculture. American Journal of Alternative Agriculture 2(4):148–152.

Edwards, C. A. 1989. The importance of integration in sustainable agricultural systems. Ecosystems and Environment 21:25–35.

Edwards, C. A., R. Lal, P. Madden, R. H. Miller, and G. House, eds. 1990. Sustainable Agricultural Systems. Ankeny, Iowa: Soil Conservation Society of America.

Grove, T. L., C. A. Edwards, R. R. Harwood, and C. J. Pierce Colfer. 1990. The Role of Agroecology and Integrated Farming Systems in Agricultural Sustainability. Paper prepared for the Forum on Sustainable Agriculture and Natural Resource Management, November 13–16, 1990, National Research Council, Washington, D.C.

Houghton, R. A. 1990. The future role of tropical forests in affecting the carbon dioxide concentration of the atmosphere. Amnio 19(4):204–209.

Jodha, N. S. 1990. Sustainable mountain agriculture: Some predictions. Paper prepared for the Forum on Sustainable Agriculture and Natural Resource Management, November 13–16, 1990, National Research Council, Washington, D.C.

Lal, R. 1986. Conversion of tropical rainforest: Agronomic potential and ecological consequences. Advanced Agronomy 39:173–264.

Lal, R. 1988. Soil degradation and the future of agriculture in sub-Saharan Africa. Journal of Soil and Water Conservation 43(6):444–451.

McNeely, J. A. 1988. Economics and Biological Diversity: Developing and Using Economic Incentives to Conserve Biological Resources. Gland, Switzerland: International Union for the Conservation of Nature and Natural Resources.

Melcher, J., and P. Stanbury. 1990. Approaches to Interdisciplinary Research: Recommendations for the SANREM CRSP. Paper prepared for the Panel for Collaborative Research Support for AID's Sustainable Agriculture and Natural Resource Management Program, National Research Council, Washington, D.C.

Myers, N. 1980. The Conversion of Tropical Moist Forests. Washington, D.C.: National Academy of Sciences.

Myers, N. 1989. DeForestation Rates in the Tropical Forests and Their Climatic Implications. London: Friends of the Earth.

National Research Council. 1984. Environmental Change in the West African Sahel. Washington, D.C.: National Academy Press.

National Research Council. 1989a. Alternative Agriculture. Washington, D.C.: National Academy Press.

National Research Council. 1989b. Investing in Research: A Proposal to Strengthen the Agricultural, Food, and Environmental System. Washington, D.C.: National Academy Press.

National Science Board. 1990. Loss of Biological Diversity: A Global Crisis Requiring International Solutions. NSB-89-171. Washington, D.C.: National Science Foundation.

Office of Technology Assessment. 1987. Technologies to Maintain Biological Diversity. OTA-F-330. Washington, D.C.: U.S. Government Printing Office.

Pimentel, D., J. Allen, A. Beers, L. Guinand, R. Linder, P. McLaughlin, B. Meer, D. Musonda, D. Perdue, S. Poisson, S. Siebert, K. Stoner, R. Salaziar, and A. Hawkins. 1987. World agriculture and soil erosion. Bioscience 37:277–283.

Ruttan, V. 1988. Sustainability is not enough. American Journal of Alternative Agriculture 3:28–130.

Ruttan, V. 1989. Biological and Technical Constraints on Crop and Animal Productivity: Report on a Dialogue. Staff Paper P89-45. St. Paul: University of Minnesota Institute for Agriculture, Forestry, and Home Economics.

U.S. Environmental Protection Agency. 1990. Greenhouse Gas Emissions from Agricultural Systems: Summary Report. Washington, D.C.: U.S. Environmental Protection Agency.

Wilson, E. O., ed. 1988. Biodiversity. Washington, D.C.: National Academy Press.

Yohe, J. M., P. Barnes-McConnell, H. Egna, J. Rowntree, J. Oxley, R. Hanson, and A. Kirksey. 1990. The Collaborative Research Programs (CRSPs): 1978 to 1990. Paper prepared for the Forum on Sustainable Agriculture and Natural Resource Management, November 13–16, 1990, National Research Council, Washington, D.C.

APPENDIX A

Introduction to Operational Issues

David Bathrick

In 1989, the Congress of the United States responded to growing environmental concerns with important legislation concerning sustainable agriculture, including legislation that mandates new initiatives within the Agency for International Development (AID). The congressional directive to AID is two-pronged: augment the current programs of the Office of Agriculture by emphasizing sustainable agriculture, and undertake a new activity that focuses specifically on sustainable agriculture. The administrator of AID underlined that directive in a recent statement in which he expressed his view that addressing environmental issues is paramount in AID's mission.

Clearly, AID has reached a major crossroads. The drive for sustainable agriculture is one expression of an evolutionary process that involves a wide range of perspectives and professions. The deliberations of the National Research Council's Panel for Collaborative Research Support for AID's Sustainable Agriculture and Natural Resource Management (SANREM) program will be highly significant as AID responds to the challenges of sustainable agriculture and the important opportunities that Congress has presented. New challenges and opportunities are also reflected in the emerging views of such organizations as the Technical Advisory Committee of the Consultative Group on International Agricultural Research. The criteria used to determine the research agenda for sustainable agriculture will be crucial to the process, as will the development of mechanisms by which the particular disciplines can cooperate and concentrate on priorities for the future.

David Bathrick is the director of the Office of Agriculture, Agency for International Development, Washington, D.C.

Since the late 1970s, AID's Collaborative Research Support Programs (CRSPs) have been bringing together some of "the best and brightest" to address agricultural problems. Recognizing this record of achievement, Congress has called for a new, sustainable agriculture CRSP, following the model employed by existing CRSPs. From the point of view of the Office of Agriculture, many of the traditional precepts of the CRSPs relate to the task at hand. Research in sustainable agriculture demands long-term institutional commitment. The congressional mandate to AID calls for a 3-year commitment, but the position of the Office of Agriculture, the Bureau for Science and Technology, and AID as a whole is that the magnitude of the task ahead will require a long-term commitment. Based on what the Office of Agriculture is hearing from AID's regional bureaus, from agriculturalists, and from the environmental community, sustainable agriculture will be the focus of even greater attention in the future, and AID needs to know more to be truly responsive. Research, then, requires the long-term commitment—longer even than the CRSPs have been around—that has been the cornerstone of the CRSP approach.

Meanwhile, the CRSPs themselves have evolved, and their work has gained increased influence on many aspects of development. The AID overseas missions, private voluntary organizations, and others with whom they are now collaborating have gained greater access to their deliberations and the results of their work. The service dimension has been added to CRSP activities, broadening them where appropriate beyond research, and university collaboration has brought matching funds—25 percent as a minimum, but in many cases close to 50 percent, to CRSP activities.

Training, mainly in agricultural disciplines, has been another basic aspect of CRSP work. The scope of training will no doubt be broadened as the multidisciplinary issues essential to the new program are addressed. Multidisciplinarity has been key to the success of the CRSP system, though in varying degrees, and the new CRSP must build on that experience to achieve true interdisciplinarity. It should look at improved technologies, but also at the processes and methodologies relevant to sustainable agriculture. Finally, the collaborative emphasis of the CRSPs, and the "win-win" relationships they foster, are highly significant. The U.S. scientific community is working with professional counterparts in the international research system, on agendas that are of mutual interest. The development of sustainable agriculture systems will clearly be of increased concern both in the United States and overseas, and CRSP research will present new opportunities for U.S. leadership.

There are no models to guide the evolution of a conceptual framework for the SANREM CRSP, and the ideas generated by the Panel for Collaborative Research Support for AID's SANREM program will be vital to the Office of Agriculture in implementing the program. It is worth noting that the CRSP system itself met with some skepticism when it was begun, but with determination, strong guidance, and prudent leadership, a viable system was created. The Office of Agriculture, the Bureau for Science and Technology, and AID's regional bureaus see the new CRSP as one key element of a sustainable agricultural program. Sustainability is the overarching theme. The overall program encompasses over 20 activities apart from the CRSPs. The management challenge for the Office of Agriculture is to determine how each program can make its particular contribution, how to augment certain facets, complement others, and fill gaps for the expanded mission.

The point has been made repeatedly that the work of the SANREM CRSP must be systems oriented, incorporate interdisciplinarity, and build on the experiences of farming systems research from a still broader range of perspectives. This has definite management implications. The ecological dimension, until now never a key in designing and managing farming systems, has to become the cornerstone. Farming systems have traditionally focused on isolated agricultural considerations; now there is growing appreciation that livestock-agriculture-soil-water-forestry linkages are fundamental parts of real sustainable systems from a farmer's perspective. Social concerns, including those of economists, must also be incorporated in a true spirit of interdisciplinarity. People must also have a sense of mutual respect as part of a committed team. This implies more than just sending messages back and forth and meeting periodically. The new CRSP must build on component research, expand it to a wider range of challenges, and integrate it at a higher level than is now done. Unfortunately, U.S. institutions—on both the AID side and the university side—do not necessarily lend themselves to this kind of collaboration. The Office of Agriculture must look closely at this aspect of the new CRSP.

This suggests another theme: decentralization at the farm level, the watershed level—whatever scale is appropriate—and going beyond the experiment station and research station structure. To do this, research must be responsive to local constraints and concerns. Certain issues are clearly identifiable as major constraints of broader concern, and as such are important in gaining local and national political support for sustainable systems. A learning process must take place, not only at the scientific level, but at the policy level, in the host country's capital. This has obvious management implications.

Matters relating to data and information management and modeling must also be considered. Clearly, yield per land unit is a very useful concept, but a much broader concept is needed. The concept of maximum sustainable yield, which is used in fisheries management, can be thought of as a tool to stimulate thinking on other approaches. A flexible spatial dimension is required, one that goes beyond the plot to farm and watershed scales. Likewise, temporal dimensions must be carefully articulated in order to take into account fallows and intercrops and their impact on productivity and the environment. These tiers of information are rarely taken into full consideration. This broader range of factors must be weighed, however, if the scientific basis of sustainability is to be determined—if sustainability is to be understood, recorded, and monitored over time.

Another management concern relates to monitoring and feedback. The manner of working within the Office of Agriculture has not often encouraged the decentralized and interdisciplinary approach that will be essential to this new thrust. These issues need to be viewed with a different perspective. Consequently, the SANREM CRSP will require leadership to mobilize and bring about monitoring of its performance in order to keep it on track and disciplined. This will require a service dimension that builds on AID's experience in working with private voluntary organizations and the for-profit private sector. Similarly, this CRSP will entail greater networking and training responsibilities, and in this it can take advantage of previous CRSP experience.

Finally, there is the matter of leveraging resources. The Office of Agriculture knows that it cannot accomplish everything with $3.3 million a year, especially as the list of interests and concerns continues to grows. Experiences such as that of the sorghum millet CRSP, which leveraged more than twice the amount of its basic funding through entrepreneurial activities and international agricultural research centers linkages, are salutary, however. Other donors, including the U.S. Department of Agriculture and U.S. foundations, will be watching this new CRSP very closely. The Congress, given its interest, will watch for short-term impacts from which broader support can be generated.

In sum, this new program will present tremendous challenges and opportunities. The Office of Agriculture has a chance to mobilize, in the true spirit of the program, the best and the brightest in the search for flexible, effective systems that can respond to the new challenges of sustainable agriculture.

APPENDIX B

Sustainable Agriculture, International Agricultural Research, and Strategies for Effective Research Collaboration

The first section of this appendix reviews the contributions of national and international organizations and U.S. universities to research on sustainable agriculture. The second section highlights the importance of increasing farmer participation in sustainable agriculture research, following which several strategies for effective collaborative research are outlined. The concluding section draws implications for the role of the Sustainable Agriculture and Natural Resources Management (SANREM) program in advancing sustainable agriculture.

INTERNATIONAL RESEARCH ON SUSTAINABLE AGRICULTURE

The Collaborative Research Support Program (CRSP) on SANREM to be established by the Agency for International Development (AID) will be an important contributor to the effort to promote sustainable agriculture and

This discussion is based on two background papers prepared for the Forum on Sustainable Agriculture and Natural Research Management held in Washington, D.C., on November 13–16, 1990: "Contributions of International Agricultural Research Centers, Agency for International Development, Food and Agriculture Organization, and U.S. Department of Agriculture to Sustainable Agriculture and Gaps in the Information Base," by Charles B. McCants, professor emeritus, North Carolina State University; and "Forging Effective Broad-Based Linkages for Sustainable Agriculture Research Among Universities, International Agricultural Research Centers, National Agricultural Research Systems, Nongovernmental Organizations, and Farmers," by Thomas B. Fricke, director, Guild for Sustainable Development, Marlboro, Vermont. Copies of these papers are available through the National Research Council's Board on Science and Technology for International Development.

natural resource management, but it will not be working alone, nor will it lack precedents to guide it in its task. On the contrary, the new program will be building on, working with, and working through other institutions that have long been committed to various aspects of sustainability. To understand better the function of the program, the institutional context in which it will operate, and the special role it can play, it is worthwhile to review how other institutions have taken on the challenge of research on sustainable agriculture.

NATIONAL AND INTERNATIONAL AGRICULTURAL RESEARCH INSTITUTIONS

The growing recognition that human welfare, environmental concerns, and development strategies are fundamentally interconnected is reflected in the greater attention that established agricultural research institutions are devoting to sustainability. Under the broad rubric of sustainability, these institutions are initiating new projects, and integrating proven ones, in the necessary effort to introduce a broader and longer term systems perspective to agriculture and development. The SANREM program itself is an important expression of this process. The established research institutions can provide, and in many cases have long provided, the professional, educational, and scientific leadership that meeting the challenge of sustainable agriculture will require. Sustainable agriculture and resource management will, virtually by definition, demand the involvement of organizations and institutions beyond those that have traditionally undertaken agricultural research. This review of the principal agricultural research institutions, and of the role of private, voluntary, and nongovernmental organizations (NGOs), highlights the important initiatives that are under way in these sectors to explore and promote sustainability—initiatives that the new CRSP must draw on in developing its own agenda.

National Agricultural Research Systems

The national agricultural research systems (NARSs) are the main mechanisms through which national governments coordinate and conduct agricultural research. Often working closely with international agricultural research centers (IARCs; described below), NARSs obtain, develop, and adapt agricultural technologies and innovations to increase productivity. In the past few years, some NARSs have begun to address sustainability issues or to develop sustainable agriculture programs. This trend can be attributed to several factors: local environmental and economic problems associated with conventional systems, the positive results of farmer participation in research, and increased demand for assistance on the part of nongovernmental organizations, farmers, and others. This trend is particularly evident in West

Africa, where NARSs lack resources and farmer participation is widely supported by nongovernmental organizations. Many other NARSs, however, continue to rely primarily on IARCs and universities for technologies, methodologies, and expertise.

The diversity of the NARSs makes it difficult to generalize about the status of their research efforts even within the principal geographic regions. The International Service for National Agricultural Research and others have reviewed the general status of the NARSs (see, for example, African Academy of Sciences, 1990; Hariri, 1990; Jain, 1990; and Senanayake, 1990). Although there are some striking examples of well-focused, mission-oriented interdisciplinary research, and even some long-term programs, that have important components that can contribute to understanding sustainable agriculture, few projects exhibit the types of integrated approach sought for the new CRSP across disciplines and institutions, and from farmer to policymaker. Many NARs, especially in Africa, lack the resources to undertake more than traditional commodity trials unless subsidized by international or bilateral donors.

International Agricultural Research Centers

The international agricultural research centers contribute significantly to the development of production systems and the technological base critical to sustainable agriculture. The IARCs were established in the 1960s and 1970s to complement NARS research on crops, commodities, and farming systems. Many of the initial centers (for example, the Centro Internacional de Mejoramiento de Maíz y Trigo, Centro Internacional de la Papa, and the International Rice Research Institute) were commodity based and focused on developing and disseminating highly productive varieties and technology packages. A second group of centers (for example, Centro International de Agricultura Tropical, International Crops Research Institute for the Semi-Arid Tropics, and International Institute of Tropical Agriculture) focused on improving multiple crops and cropping systems in specific agroecological zones or bioregions. The goal of all these centers is to improve agriculture in developing countries by using research as a tool for change. Of the 19 IARCs, 13 are members of the Consultative Group on International Agricultural Research (CGIAR), an association of countries, international and regional organizations, and private foundations that support a worldwide system of agricultural research centers and programs.

The IARCs have had a major impact on agricultural production through the development of improved varieties of major crops (notably wheat, rice, and maize) that are grown in developing countries. The research efforts of the IARCs have also proven effective in improving the efficiency of water and fertilizer use, strengthening the overall performance of farming systems, and training developing country scientists. The CGIAR now recognizes that agricultural sustainability is a dynamic challenge that must be met by developing countries within the context of rapidly growing demands for food and fiber. The CGIAR has identified the following four major sustainability concerns that will provide the focus for its research (Consultative Group on International Agricultural Research, 1990).

• Protection of the genetic base for agriculture. For the commodity-based research centers, this is considered to be a primary task. As improved cultivars are grown over wider ecosystems, a broader range of resistance to pests and ability to respond to environmental constraints become paramount. Maintaining the genetic base of future cultivars through protection and preservation of natural diversity thus becomes increasingly important.

• Preserving the natural resource base. Although they recognize the importance of cultivars in achieving sustainable agriculture, the IARCs are aware of the critical role of the natural resource base. Accordingly, they have increased their efforts to ensure that current production methods do not undermine the ability of future generations to meet their natural resource needs. A central theme of the IARCs' mission is to design agricultural systems that do not force a trade-off between current and future production, that is, systems that are sustainable even as they meet expanding production needs. The IARCs recognize that sustainability of agricultural production hinges on improved efforts to manage natural resources. Increasingly, they are recognizing the need for policies and programs that encourage soil and water conservation, long-term investments in improving common property resources, and the application of new techniques, such as conservation tillage.

• Problems pertinent to less favorable environments. Most of the IARCs are giving increased priority to, and allocating more resources in support of, research that can help to resolve production and sustainability issues in agroecological settings where stress conditions are dominant. The current research strategies can be divided into two categories: (a) "research to raise and sustain output in high production systems in favorable environments such as irrigated areas or fertile rainfed areas" and (b) "research to meet the needs of farmers in areas where production is constrained by unfavorable agroecological conditions" (Consultative Group on International Agricultural Research, 1990). The increased emphasis on research in the second category is especially prominent in those IARCs established to serve marginal ecosystems (for example, the International Center for Agricultural Research in the Dry Areas and the International Crops Research Institute for the Semi-Arid Tropics) and among nonaffiliated centers that were established to study particular production factors (for example, the International Board for Soil Research and Management and the Indian Council on Agricultural Research).

A further element is being added to research in those IARCs that have historically focused on increased yields of major cereals in irrigated and other favorable environments. For example, the research strategy of the International Rice Research Institute has evolved from one focused primarily on increasing aggregate rice production to one that balances regional production growth with the needs of poor farmers who depend on marginal ecosystems. The challenges of achieving sustainability in fragile areas are substantially different from those encountered in more intensive agricultural areas. It is unlikely that major breakthroughs will occur in any one crop, farming system, or input package that will result in sustained yield increases in such areas. Diversification may well offer the most attractive alternative. Consequently, research is being increasingly directed toward developing genetic traits conducive to raising productivity within integrated cropping and livestock systems that offer a higher probability of sustainability in marginal environments. This research concentrates on the development of cultivars that are suitable to adverse ecological conditions and that are compatible in mixed farming systems, but it also includes work on efficient mixed and relay cropping systems, crop and livestock interactions, agroforestry systems, and minimum tillage.

• Sustainable agriculture and external inputs. The IARCs recognize that concerns of sustainability overlap concerns about reducing costs for poor farmers, and they have increased their efforts to develop a better understanding of how they can maximize the use of on-farm resources to increase agricultural production. Thus, more attention is being given to biological and ecological interactions, nutrient-cycling techniques, and organic matter and pest management practices that require a minimum of purchased inputs. More social science research is being conducted on how to enable farmers to deal more effectively with the political and economic constraints on sound natural resource management practices. The IARCs recognize the need for greater outreach and cooperation with organizations directly concerned with natural resource management and the application of scientific and technical information to sustainable agriculture.

In the future, the CGIAR centers plan to give greater emphasis to components of sustainable agriculture and to address the more complicated, multidisciplinary issues of agroecosystem management, long-term measurement of sustainability, and interactions between technology and institutional policy (Consultative Group on International Agricultural Research, 1990). More broadly, their contributions to sustainable agriculture will likely concentrate on activities they have always done well: (a) developing genotypes that permit greater efficiency in the use of the natural resources within particular agroecological settings; (b) promoting component research that maximizes the integration of biological processes, enhances soil fertility, and protects production systems from pests and nonbiological stresses; (c) designing technologies that do not force trade-offs between current and future production systems and that sustain or enhance the natural resource base; (d) undertaking socioeconomic studies that will help make sustainable systems more acceptable; and (e) assisting national agricultural research systems—through cooperative research, training, and information exchange—in contributing to and creating conditions for national sustainable agricultural development.

FOOD AND AGRICULTURE ORGANIZATION OF THE UNITED NATIONS

Recently, the Food and Agriculture Organization of the United Nations has undertaken a number of initiatives that will have a major influence on its future programs on the environment and sustainable development. These include the formulation of a strategy and action agenda for sustainable agricultural development, attention to plant genetic resource issues relative to biodiversity and biotechnology, and consideration of the effects of climatic change on agriculture, forestry, and fisheries (Food and Agriculture Organization, 1990).

U.S. DEPARTMENT OF AGRICULTURE

Sustainable agriculture is a major theme within the U.S. Department of Agriculture's Agricultural Research Service. Recently, the Agricultural Research Service reviewed current research activities considered to be supportive of sustainable agriculture and placed them in the following categories (U.S. Department of Agriculture, 1989):

• biological pest control and integrated pest management of insects, soil nematodes, crop toxins, weeds, and internal animal parasites and diseases;
• improvement of crop varieties for resistance to acid soils, air pollution, insects, soil nematodes, diseases, drought, and other stresses;
• water and soil management to conserve water, improve water quality, and sustain production;
• management systems that are economical, environmentally sound, and sustainable;
• erosion control;
• nutrient management to reduce fertilizer use, avoid water pollution, and maintain yields;
• forage production and animal production; and
• beneficial organisms.

Although this research seeks to develop information useful to guiding U.S. agriculture toward sustainability, the basic principles that issue from it can also be valuable in formulating management practices in countries throughout the developing world. Alternative pest control methods; efficient soil, water, and nutrient management practices; and the development of crop varieties that produce economical yields under stress conditions—all are fundamental inputs for any sustainable agriculture system.

A recent addition to the sustainable agriculture initiative is the special program originally referred to as LISA (low-input sustainable agriculture), now simply called sustainable agriculture. This approach to farming uses lower amounts of purchased inputs, such as fertilizers and pesticides, and emphasizes greater reliance on on-farm resources and naturally occurring processes (U.S. Department of Agriculture, 1990a). It gives considerable attention to efficient use of natural resources and environmental protection, but it also emphasizes profitability, based on the premise that a farming method must be profitable to be sustainable. Because the concept calls for the careful integration of various components in the production scheme, its effective implementation requires skilled and intensive management. Low-input practices vary from farm to farm, but they commonly emphasize the use of crop rotations, soil and water conservation, crop and livestock diversification, mechanical cultivation, animal and green manures, and biological pest control (U.S. Department of Agriculture, 1990b).

AGENCY FOR INTERNATIONAL DEVELOPMENT

Making agriculture more sustainable has been an implied, if not expressed, goal of AID since its inception. In support of this objective, AID has provided leadership and funding for a wide range of research and development programs that have made important contributions to the cause of sustainability.

The Agency for International Development has been a primary source of funds for the core budgets of the IARCs—from a high of approximately 30 percent to the current 18 percent of the total core budgets. This commitment has given the IARCs the financial stability essential for long-term research on sustainable management practices. Examples include the development of cultivars adaptable to stress conditions, soil and water management practices that enhance plant growth, and farm management systems that minimize erosion and environmental degradation.

The Agency for International Development has also been the primary source of funds for U.S. university and other nongovernmental research and development programs tailored to the unique needs of the developing world. This support has given such institutions the opportunity to develop their high level of expertise in international development and can contribute new technologies in the design of sustainable agricultural systems. The CRSPs (collaborative programs involving AID, U.S. universities, and host country institutions) are a primary component of this effort, and they are making major strides toward sustainable agriculture. Although total costs are shared among the collaborators, the CRSPs depend heavily on AID for operational funds.

In developing countries, AID provides leadership and financial support to enable local agencies to address policy, institutional, and operational issues influencing sustainable agriculture programs. Actions that lead to available credit, ready markets, and stable land tenure are as important to the realization of sustainable agriculture as are improved cultivars or efficient soil management practices.

The support from AID has come from many of its structural units. Regional bureaus have focused on issues of highest priority within the countries of their respective regions. Specific activities have included institution building, policy reforms, and technology adaptation and promotion—all of which are essential to sustainable agriculture. The Bureau for Science and Technology has primary responsibility for providing leadership in the development of new science-based technologies within AID. Recently, the bureau has been directed to place additional emphasis on actions needed to support and promote sustainable agriculture. Within the bureau, the primary leadership is expected to come from the Office of Forestry, Environment, and Natural Resources, the Office of Agriculture, and the Office of Rural and Institutional Development. The focus of the Office of Rural and Institutional Development is on cross-cutting institutional and human issues that can broaden people's economic opportunities and sustain economic growth (Agency for International Development, 1990a,b). Increasing people's access to production resources and technologies broadens their economic opportunities.

A major underlying cause of natural resource degradation in the developing world is human poverty. Thus, efforts to achieve sustainability must encompass increased economic returns as well as increased productivity (Agency for International Development, 1987). A part of the program strategy of the Bureau for Science and Technology's Directorate for Human Resources is based on the premise that sustainable natural resource utilization is achieved by strengthening those human incentives and institutions that encourage rational use of natural resources critical to economic growth. In support of this strategy, the directorate supports a range of programs that address land tenure and access issues, human and institutional factors related to the use of multipurpose tree species, and the expansion and refinement of geographic information systems. The following are examples of directorate projects that contribute to the goal of sustainable agriculture (Agency for International Development, 1990b):

• Access to land, water, and other natural resources. The purposes of this program are to improve understanding of the relationships between resource tenure and sustainable growth and to facilitate the application of such understanding to development programs and policies.

• Development strategies for fragile lands. This program assists cooperating countries in developing and implementing strategies to arrest degradation of fragile lands so as to foster sustained production of food, fuel, and income.

• Forest/fuelwood research and development. This project supports research and management activities that promote country-specific fuelwood and forestry plans and programs.

The activities of the Bureau for Science and Technology's Office of Agriculture are guided by the AID goals of increasing incomes of the poor, expanding the availability of food, and maintaining and enhancing the natural resource base. The office supports research and development involving crops, livestock, fisheries, soil, water, economics, and agricultural policies. It has a large and diverse group of projects, most of which contribute to components or processes essential for sustainable agriculture.

The Office of Agriculture also manages AID's technical and scientific relationships with the international agriculture research centers within the CGIAR and with the International Fertilizer Development Center, a nonaffiliated center (Agency for International Development, 1990c). The goal of the latter is to ensure that farmers in developing countries have a dependable and economical supply of fertilizers to meet special crop and soil requirements. It conducts research on fertilizer use to provide guidance in the selection of rates, methods of application, and sources that will relieve nutrient constraints to plant growth in an agronomically and environmentally sound manner. In addition, all of the CRSPs are funded through and managed by the Office of Agriculture. The office also supports other collaborative research support projects, primarily with U.S. land-grant universities, that are highly relevant to sustainable agriculture (Agency for International Development, 1990c). These include the following:

• improving cropping systems through the use of soil-improving legumes,
• developing technology for soil moisture management,
• using biotechnology to improve animal vaccines,
• developing models and expert systems to evaluate options for sustainability in agriculture,
• improving biological nitrogen fixation, and
• increasing knowledge and understanding on how economic policies affect agricultural development.

In its recently released strategic plan for the 1990s, the Office of Agriculture emphasizes that agricultural development in the developing countries is a paramount component of sound economic growth (Agency for International Development, 1990c). The ability to continue agricultural growth by expanding land under cultivation is no longer a viable option. The challenge for the future is to develop cost-reducing technologies and appropriate policies that can increase yield per unit of area and time while maintaining the natural resource base. This goal must be achieved through science-based practices that are economically remunerative, environmentally sound, and organized within a coordinated framework that is based on the concepts of sustainability.

U.S. UNIVERSITIES

The U.S. universities, and the academic sector in general, are involved in international research on sustainable agriculture through many different programs (including those listed above), in many different regions, and with many different emphases. Many of the land-grant universities have been actively involved for decades in sustainable agriculture research through their colleges of agriculture, but research efforts germane to sustainability issues are just as likely to be found in universities and university departments that have not traditionally been involved in basic agricultural research (for example, geography and anthropology departments, regional studies programs, environmental focus programs, and interdisciplinary institutes). These other academic sectors, which may or may not be active in current collaborative research support programs, must be identified and engaged if the SANREM program is to be truly innovative, interdisciplinary, and effective.

In recent years, more and more universities have been expressing support for and targeting their resources to sustainable agriculture research. Often this research is conducted within traditional departments. In other cases, special departments, centers, and/or programs have been created to stimulate the necessary interdisciplinary efforts this research requires. This is the case in such major land-grant universities as Ohio State University, the University of California at Santa Cruz, the University of Maine, the University of Wisconsin, and Iowa State University. Many of these programs are new and still gaining their institutional "wings"; the SANREM program may, in this instance, provide additional benefits by helping to strengthen these fledgling initiatives. In addition, the academic and professional societies that serve to unite these diverse programs and institutions are an important channel for gathering and disseminating information within the rapidly growing community of researchers—traditional agricultural researchers as well as scholars and scientists from other fields—who are studying various aspects of sustainable agriculture and natural resource management.

A review of major institutions active in research on sustainable agriculture leads to several conclusions. First, much is being done that is contributing to the information base of sustainable agriculture. Second, much is being planned for future programs that will broaden as well as deepen this information base. And, third, the level of commitment to these programs and to sustainability as a goal is and must remain high over time to ensure progress.

Research on sustainable agriculture contains few if any absolute voids, but in some areas the information base is thin and the level of effort lower than their importance would seem to justify. Three areas in particular deserve greater attention: synthesis and analysis of information, improvement in the knowledge base, and strategic planning.

Synthesis and Analysis of Information

The limitation on success in solving many situation-specific problems is not so much what is not known as the inability to use what is known. Much of the relevant information has not been organized, integrated, or delivered, and thus it is not available to guide decisions. The effective information base contains scientifically ascertained data as well as personal insight and experience. Some progress in organizing this vast body of knowledge has been made through the development of simulation models, systems, and other mechanisms that take advantage of the powerful advances in computer technology. Although these efforts show great promise, they must be applied more broadly in making on-site management decisions. Greater cooperation in identifying information gaps and planning experiments would help to bring this about.

Improvement in the Knowledge Base

With respect to the knowledge base, two general areas need more work. First, researchers must characterize landscapes with respect to carrying capacity (both human populations and biodiversity more generally), production potential, development constraints, and risks—and do so in a manner that provides guidance for achieving sustainable agriculture in that area. In evaluating these features, researchers must not only identify natural characteristics of the landscape, but also provide information on the limits to sustainable agriculture in the given area and the level of risks to be expected when the development process is undertaken. Second, researchers must take greater advantage of indigenous knowledge in identifying constraints to development and in prioritizing development efforts in specific situations. This approach is unlikely to reveal heretofore overlooked miracle solutions to fundamental problems, but it will provide invaluable leads that can aid in meeting the requirements of local farmers and local environments.

Strategic Planning

In terms of strategic planning for advancing sustainable agriculture, two needs stand out. First, more emphasis must be given to interdisciplinary research. Many of agriculture's most pressing problems demand a systems approach and the interaction of different disciplines. If more interaction takes place at the front end of the research process—as opposed to belated attempts to fit the pieces together after research on individual components is completed—the effectiveness of the work as a whole will be improved. The call for interdisciplinary research is not new, and success in this regard will not be overwhelming. Nonetheless, this is a highly significant aspect of research on sustainable agriculture, and as such it warrants continued and constant encouragement. Second, mechanisms are needed that can provide management responses to specific, on-the-ground problems and to the conditions that fostered them. Each of the many determinants of sustainability entails a broad range of variables and interactions. In combination, they create an essentially infinite number of conditions that can and do occur. Each farmer or producer has unique circumstances and needs; broad generalizations or recitations of fundamental principles of soil, crop, and livestock management are therefore seldom useful. Providing management options that are based on sound, complete information and that are communicated in an understandable manner will always be the most practical way to fit proper decisions to individual circumstances.

THE CRITICAL CHALLENGE: REACHING THE FARMER

The above review of agricultural research institutions suggests not only the desirability of, but also the need for and effectiveness of, broad-based collaboration in sustainable agriculture research. Many of the CRSPs' research efforts have clearly defined goals, roles, and structures and function well. Their success reflects the ability to adapt to changing needs and opportunities.

Ineffective collaborative efforts, however, can also be found. If research objectives are unfocused or overly ambitious, research projects can lose momentum or collapse. If linkages engage incompatible entities, collaboration can be difficult or impossible. Moreover, poorly conceived and managed collaboration can be very costly and unproductive and can undermine chances for further collaboration in the country or research area. Constituencies may work against each other for good reasons, and barriers to communication are often formidable. For example, differing objectives and approaches among NGOs, farmers, research institutions, and other participating groups may inhibit collaboration. Disparities in the allocation of resources and spheres of influence may further inhibit collaboration. These barriers must be overcome if the SANREM program is to succeed in working with the farmer to identify constraints, conduct research, and disseminate the knowledge and tools that promote sustainability.

Nongovernmental organizations are playing an increasingly important role in reaching farmers, particularly in developing nations. The private organizations that make up the NGO community are highly heterogeneous. They may be international or indigenous, community-based or national associations, rural farmers as well as technical and financial support intermediaries, and networks for information dissemination and for cross-cultural exchange. In general, the number of NGOs with extensive research and extension capabilities is still small. In the United States, a small group of NGOs, including Rodale International, CARE, World Neighbors, and Winrock International, have well-developed field capabilities in sustainable agriculture. Until recently, NGO linkages with IARCs and NARSs have been limited by mutual distrust or a lack of collaborative mechanisms.

Many NGOs have strong farmer outreach capacities. They strive to create horizontal rather than vertical linkages with farmers to stimulate agricultural improvements and innovations. Most NGOs actively engage farmers in an attempt to reduce the gap between basic and applied research. Although relatively limited in terms of technical resources and scientific rigor, NGOs, with their emphasis on field-based approaches, serve as increasingly critical links between farmers and scientists. They can also play useful roles in shaping policy and the research agenda. National and regional NGO networks and agencies are able to articulate and advocate research priorities. Organizations that exemplify these roles include the Committee on Agricultural Sustainability in the United States, the Sustainable Agriculture Coalition in the Philippines, and the Latin American Consortium on Agroecology and Development in Latin America.

Until the emergence of the sustainable agriculture concept, the various constituencies tended to polarize. Parts of the NGO community regarded the IARCs, NARSs, and university faculties as promoters of top-down cropping systems and technologies that were unsustainable, commodity based, and high in chemical and energy input. For many NGOs, such systems ignored rural poverty and environmental degradation. Conversely, parts of the research community viewed NGOs and farmers as strident, unscientific, or naive.

Because it emphasizes and incorporates interdependencies, sustainable agriculture creates conditions in which broad-based collaboration in research is not merely possible, but necessary. The successful examples of collaboration—and most concerted attempts are successful—show how the various constituencies have created effective research linkages. There are no precise formulas or shortcuts, however. The literature describing research collaborations is generally project specific or overly conceptual. The system of collaborative linkages has been developed largely by organizational development experts. Their work emphasizes strategic management, team building, communications, conflict resolution, and learning processes. Information from the practical side is limited. Only a few field practitioners and program managers have published useful reflections on their experiences (see, for example, Agency for International Development, 1984; Brown, 1990; Osborne, 1990).

STRATEGIES FOR EFFECTIVE RESEARCH COLLABORATIONS

In looking ahead to the implementation of the SANREM program, it is worthwhile to note that the mechanisms by which collaboration occur are many and diverse and must be tailored to specific situations. It will be the responsibility of researchers to discover the mechanisms that are most suitable for the specific groups involved and research undertaken. Although authoritative case studies and standardized guidelines are generally lacking, a number of conditions and criteria that appear fundamental to successful and viable agricultural research collaboration can be identified.

Build Consensus Through Outreach and Consultation

Collaboration grows from a process of dialogue and negotiation. The various constituents must first be aware of the existence and resources of one another, be they NGOs, local farmers, or scientists. Targeted outreach must follow. Subsequently, it is important to create forums for the exchange of views and discussion of opportunities through consultations. These may be held in the field, in seminars, or at workshops. This process of collaborative consensus building requires patience and compromise. Expert facilitation and considerable personal initiative are often essential to the process.

As an example, the On-Farm Seed Project used an extensive process of outreach and consultation to develop its program and mobilize its constituencies in the United States and Africa. The initial concept emerged in the United States, through the joint initiative of Winrock International and the Center for PVO/University Collaboration in Development. The center tapped its existing network of academic institutions and development-related NGOs to discuss needs and devise strategies for improving on-farm seed technologies in West Africa. In Senegal and The Gambia, the project engaged a broad array of representatives of the national government, the Peace Corps, and local and international NGOs. The design and planning process included national consultations, on-site field needs assessments, and institutional surveys. According to project personnel, the investment in a thorough and inclusive consultative process paid off in the form of a more dynamic and effective program (Winrock International, 1990).

Establish Genuine Partnerships

Collaboration must be based not on dominance or methodological bias, but on mutual respect, partnership, and goodwill. Effective partnerships recognize complementary roles and mutual self-interest. Participants from disparate disciplines and constituencies must overcome or compromise their parochial concerns. Most successful partnerships rely on good personal relationships and compatible approaches. In the case of SANREM research, for instance, it will be important for researchers, NGO personnel, and farmers to understand and reconcile each other's views of traditional versus conventional agriculture, external inputs, and sustainability.

Rodale International's program in Senegal exemplifies the concept of research partnerships. In this program, farmers are brought into the research and extension process as equal partners. According to Rodale, the national agricultural research system in Senegal has the technical and financial resources to generate viable alternative agricultural practices. However, researchers seldom ask the right questions and are weak in developing partnerships with small farmers. The NGOs in Senegal have strong links with farmers, but they lack the capacity to do applied research. Thus, the NGOs act as the bridge between the national system and local communities. By working together, all of these constituencies greatly enhance their impact.

Develop Shared Objectives, Work Plans, and Time Frames

Research objectives and work plans should be demand driven and field based to the fullest extent possible. All parties should participate in program planning and design in the field or as close to the research sites as possible. Time frames should be based on realistic expectations and should take into account the opportunity costs for each party's time and labor. Finally, logistical and financial constraints should be discussed and reconciled prior to starting the research.

This principle is well illustrated by two research and technology development initiatives in West Africa: the Farmer Innovations and Technology Testing Project in The Gambia (Gilbert, 1990) and the Low-Resource Agriculture Project in Liberia (which has been interrupted by civil strife). In both projects, representatives of farmer groups, NGOs, and the national agricultural research system cooperated fully in the design and planning process. The teams first established a long-range research framework and assigned specific responsibilities among the national system, universities (in Liberia), NGOs, and farmer groups. The planning teams then identified the cropping systems and technology that would be the initial focus of the research, as well as testing objectives, timetables, and resource allocations. Both projects make provision for further modifications and innovations by farmers and other constituencies.

Pursue Practical and Feasible Research Agendas

Collaboration is enhanced when research agendas can generate practical, applied results within a reasonable period. Research objectives should be focused topically or geographically; they should not be overly ambitious nor dispersed. Demands for closure on objectives vary among constituencies. Farmers and NGOs, in particular, must be able to apply and disseminate research results. Thus, close linkages between research and extension is highly advantageous, for example, through inclusion of on-farm adaptive trials and demonstrations.

The two initiatives based in Senegal have created close linkages among research, extension, and broader dissemination. Both programs have attracted ongoing farmer involvement by addressing their immediate priorities. The Rodale program mentioned above combines on-farm cropping systems and soil fertility research with village-based demonstrations of practical soil conservation and livestock improvement techniques. The On-Farm Seed Project is improving the quality of the 80 to 90 percent of total seed stocks in Senegal that are selected and saved on-farm. The project uses agroecological surveys and on-farm trials, and it disseminates improved technologies widely once they are proven.

Create Responsive Communications Mechanisms

Research teams should communicate effectively. Appropriate means of exchanging information, documents, and feedback should be established before research begins and should be monitored closely thereafter. Weak infrastructures may dictate the use of creative communication alternatives. Language barriers and cultural differences between farmers and researchers will also have to be addressed to avoid misunderstandings.

The Informationcentre for Low External Input and Sustainable Agriculture (1988) asserts that the interpersonal aspects of communications are more important than the technological aspects. The methodologies for farmer participatory research incorporate many useful techniques for improving communication among farmers, field-workers, and researchers, including community appraisal, innovators' workshops, and mapping and systems diagrams (Chambers, 1989). All of the projects mentioned above use similar methods to improve the flow of information on agricultural innovations, traditional and improved technologies, and research and extension results.

Install Effective Management and Decision-Making Systems

Collaboration requires flexibility, coordination, and leadership. Management entities may assume many different forms, including secretariats, lead agencies, and consortiums. Decisions should be made with the help of consultative or coordinating bodies. These bodies should represent all parties involved and should actively engage in program review and evaluation. Effective leadership in collaborative efforts responds to the needs of the various constituencies, maintains linkages, and guides the overall progress of the research. The roles and responsibilities of all the participants should be understood and, if possible, put into written agreements.

All of the projects cited above developed effective management and decision-making systems. In the On-Farm Seed Project, for example, the consortium established parallel structures in the United States and in the field. To coordinate activities, it also established a central secretariat at the headquarters of the designated lead agency in the United States. Other member agencies were encouraged or designated to assume supportive leadership roles in countries where field operations were taking place. Primary planning and coordination of program activities are conducted by a coordinating committee representing all members in the field, and an advisory board in the United States provides oversight of agendas, policies, and programs. To date, these structures are working to provide coordination, mediate conflicts, and sustain active broad-based participation.

CONCLUSION: IMPLICATIONS FOR THE SANREM PROGRAM

The collaborative linkages essential to the success of the SANREM research support program will be more difficult to establish than those associated with the existing CRSPs. The SANREM program will have to develop collaboration along multiconstituency as well as multidisciplinary lines. The examples presented above, however, illustrate that the program can achieve its objectives.

Sustainable agriculture is not an event that occurs at some point in time or at some site on the global landscape. Rather, it is a goal that is useful in focusing programs and arranging the application of resources.

This review of the activities, strategies, and plans of institutions engaged in agricultural research and development reveals clearly that much has been done, is being done, and will be done to contribute to achieving sustainability in agriculture. Much of the current work is targeted toward the components and principal determinants of sustainability; as this work continues, it will broaden and deepen the information base. The weakest point in the overall approach is the ability to integrate these components and develop management options for specific situations and conditions. The SANREM program cannot fill this gap on its own, but it can serve an important role in bringing together the wide variety of disciplines, constituencies, and organizations that must work together to put agriculture on a sustainable path.

In the final analysis, it is the individual farmer who will examine his or her resources, needs, and dreams and make the decision whether to slash and burn another hectare or to grow another crop on existing land. It is the farmer who will elect whether to use manures or purchase fertilizers, to plant a monoculture or interplant crops. It is the farmer who will choose whether to control pests by purchased inputs or by other means. Finally, it is the farmer who will decide what risk he or she is willing to take in making each of these and other decisions. The better the information at the farmer's disposal, and the easier that information is to understand and act on, the higher the probability that the ultimate decision will increase productivity while conserving agricultural resources. The thriving interest in sustainable agriculture and the increased commitment to research on its essential components and their interactions should increase the long-term benefits of these incremental decisions to the farmer, consumers, society at large, and the environment that contains and supports them all. This is not to indicate that all farmers will benefit from this approach. Sustained agricultural growth will place increased demands on farmers, and some will find the demands more than their circumstances can tolerate. Rural urban drift, however, should occur through planned and informed political choice, not by default, and to the extent possible in a manner that enables good farmers to continue to sustain agriculture, while their less-skilled or motivated brethren move into industrial and service employment, rather than the reverse.

REFERENCES

African Academy of Sciences. 1990. Scientific Institution Building in Africa: Report of the Bellagio Symposium. Nairobi, Kenya: African Academy of Sciences.

Agency for International Development and U.S. Peace Corps. 1984. A Guide to AID/Peace Corps/PVO Collaborative Programming. Washington, D.C.: Agency for International Development.

Agency for International Development. 1990a. Global Research for Sustainable Food Production. Collaborative Research Support Program Council. Bureau for Science and Technology. Washington, D.C.: Agency for International Development.

Agency for International Development. 1990b. Portfolio Directory. Directorate for Human Resources, Bureau for Science and Technology. Washington, D.C.: Agency for International Development.

Agency for International Development. 1990c. Strategic Plan of the Office of Agriculture in the 1990s. Office of Agriculture, Bureau for Science and Technology. Washington, D.C.: Agency for International Development.

Brown, M. 1990. PVO/NGO-NRMS project: Lessons learned. NRMS Newsletter (August 1990). Washington, D.C.: Energy/Development International.

Chambers, R., A. Pacey, and, L. A. Thrupp. 1989. Farmer First: Farmer Innovation and Agricultural Research. London, England: Intermediate Technology Publications.

Committee on Agricultural Sustainability for Developing Countries. 1987. The Transition to Sustainable Agriculture: An Agenda for AID. Committee on Agricultural Sustainability for Developing Countries. Washington, D.C.: Committee on Agricultural Sustainability for Developing Countries.

Consultative Group on International Agricultural Research. 1990. Sustainable Agricultural Production: Final Report of the CGIAR Committee. Document No. MT/90/18. Presented at CGIAR Meeting, May 21–25, The Hague, Netherlands.

Food and Agriculture Organization (FAO). 1990. FAO Activities Related to Environment and Sustainable Development. Report of the 98th session of the FAO Council. Rome, Italy: Food and Agriculture Organization of the United Nations.

Gilbert, E. 1990. Non-Governmental Organizations and Agricultural Research: The Experience of The Gambia. The Gambia: Gambia Agricultural Research and Development.

Hariri, G. 1990. Organization and Structure of Arab National Agricultural Research Systems (NARS). ISNAR Working Paper No. 31. The Hague, Netherlands: International Service for National Agricultural Research.

Informationcentre for Low External Input and Sustainable Agriculture (ILEIA). 1988. Participative technology development. Newsletter for Low External Input and Sustainable Agriculture (October 1988). Leusden, Netherlands.

Jain, H. K. 1990. Organization and Management of Agricultural Research in Sub-Saharan Africa: Recent Experience and Future Direction. ISNAR Working Paper No. 33. The Hague, Netherlands: International Service for National Agricultural Research.

Osborne, T. 1990. Multi-Institutional Approaches to Participatory Technology Development: A Case Study from Senegal. Dakar, Senegal: Winrock International.

Senanayake, Y. D. 1990. Overview of the Organization and Structure of National Agricultural Research Systems in Asia. ISNAR Working Paper No. 32. The Hague, Netherlands: International Service for National Agricultural Research.

U.S. Department of Agriculture. 1989. Agricultural Research, October 1989.

U.S. Department of Agriculture. 1990a. Low-Input Sustainable Agriculture (LISA). Research and Education Program of the U.S. Department of Agriculture. Cooperative State Research Service. Washington, D.C.: United States Department of Agriculture.

U.S. Department of Agriculture. 1990b. Alternative Opportunities for U.S. Farmers. Cooperative State Research Service, Special Projects and Programs Systems. Washington, D.C.: U.S. Department of Agriculture.

Winrock International and Center for PVO/University Collaboration in Development. 1990. On-Farm Seed Project. Third Annual Report. Morriltown, Ark.: Winrock International.

APPENDIX C

Soil Research for Agricultural Sustainability in the Tropics

Rattan Lal

To date, much of the increase in food production in developing countries has been achieved by bringing new land under production, expanding irrigated land area, and applying Green Revolution technologies. These means have been used to the limit as unprecedented demographic pressure has generated rapidly growing demand for agricultural products.

Reserves of potentially arable prime agricultural land are limited and unevenly distributed. The population in large areas of Africa, Asia, and South America already exceeds the carrying capacity of the land. Land is indeed a scarce resource; globally, arable land per capita will progressively decline from about 0.3 hectare (ha) currently to 0.23 ha in 2000, 0.15 ha in 2050, and 0.14 ha by the year 2150. The potentially arable land that exists, moreover, is located in regions with weak logistics, poor accessibility, and very poor infrastructure. Densely populated Asia, with up to 75 percent of the world's population, has little additional arable land to convert to agricultural use (for example, Sumatra). The per capita arable land area in many Asian countries is already less than 0.1 ha. About 290 million ha of land may be suitable for agriculture in South America and 340 million ha in Africa (Buringh, 1981; Dudal, 1982). Most of these lands, however, are located in fragile and ecologically sensitive regions—tropical rain forests, acid savannahs, the drought-prone Sahel. Bringing new land under production through deforestation of tropical rain forests has severe ecological, environmental, and sociopolitical implications. Some of the potentially arable land

Rattan Lal is associate professor of soil physics at the Department of Agronomy, The Ohio State University.

outside the tropical rain forest region is of marginal utility due to other constraints—the land is too steep, the region contains too little or too much water, and the soils are too shallow or show salt and nutrient imbalances.

Irrigation has played a major role in increasing food production. For the decade ending in 1987, the rate of increase in irrigated land area was 1.0 percent in Asia, 1.3 percent in Central and South America, and 1.4 percent in Africa (Food and Agriculture Organization, 1986). The rate of expansion has slowed considerably as the availability of irrigable land and good quality irrigation water has become a severe constraint.

Concern is growing that the impact of green revolution technologies is slowing, even in South Asia (Herdt, 1988). The influx of high-energy techniques into agricultural ecosystems has broken the yield barriers, increased output at the rate of about 2.5 percent a year, led to an overall increase in per capita food production of about 0.6 percent between 1950 and 1986, and resulted in an unprecedented boom in agricultural output in the post-World War II era. Green revolution technologies have been applied to prime agricultural land with input-responsive soils. Can this technology be applied to the impoverished soils of the humid and subhumid tropics of Africa and South America? A principal constraint may be nonavailability of essential inputs at affordable prices, the breakdown of resistance of improved cultivars to pests and pathogens, and the degradation of soil quality.

Neglect, misuse, and mismanagement of soil resources are in large part responsible for the low yields, widespread poverty, and severe problems of soil and environmental degradation in tropical and subtropical regions. Consequently, the goals of a viable program in soil research for agricultural sustainability in the tropics must be to (a) maintain and enhance the biological and ecological integrity of soil resources; (b) increase agricultural production; (c) improve the income, buying capacity, and self-reliance of resource-poor farmers; (d) restore life-support processes and potential productivity of degraded ecosystems; and (e) provide support to national research institutions and development services. The next section discusses a number of issues related to soil research for sustainable agriculture.

SOIL RESEARCH ISSUES

The severe scarcity of arable land and mounting demographic pressures in many developing countries mean that technological innovations are needed that can bring about a quantum leap in agricultural productivity. This can only be achieved through science-based agriculture. Given resource-based agriculture with low or medium input, the minimum dietary requirement can only be met with a per capita land availability of 0.5 ha. Thus, the greatest challenge facing humanity in the twenty-first century will be to produce the basic necessities of food, feed, fiber, fuel, and raw materials

TABLE C-1 Yield in Grain Equivalents and Percentage of Cropland for Various Levels of Production Input in the World

Farming System/ Input Level	Yield (kg/ha)	Cropland (%)	Average Area of Arable Land Needed (ha/capita)
Shifting cultivation	<100	2	2.65
Low traditional	800	28	1.20
Moderate traditional	1,200	35	0.60
Improved traditional	2,000	10	0.17
Moderate technological	3,000	10	0.11
High technological	5,000	10	0.08
Specialized technological	7,000	5	0.05

SOURCE: P. Buringh. 1981. An Assessment of Losses and Degradation of Productive Agricultural Land in the World. FAO Workshop on Group Soils Policy. Rome, Italy: Food and Agriculture Organization of the United Nations.

from the maximum per capita land availability of 0.14 ha or less. Technological options for sustainable management of soil and water resources in the twenty-first century must address this basic constraint.

The per capita land requirement to meet basic needs depends on the inputs. The challenge is to intensify use of prime agricultural lands, with all the inputs needed to sustain productivity of soil and water resources, and to break the yield barriers. Buringh (1981) presents an optimistic scenario. He estimated various modes of agriculture and the per capita land requirement for each mode (Table C-1). The average per capita land requirement under different systems and the corresponding crop yields vary by several orders of magnitude. Two of the most encouraging aspects of this analysis are that (a) the per capita arable land area can be as low as 0.11 ha or less for a moderate level of technological inputs and (b) about 25 percent of the world's cropland is suitable for intensive use through adoption of moderate, high, or specialized technologies.

There are other optimists who support Buringh's analysis and argue that the earth's natural resources have the capacity to support between 15 to 22 billion inhabitants (Calvin, 1986, cited in Hudson, 1989). Their estimates are premised on (a) total solar energy input on arable land and (b) grain production with improved technologies. They argue that food production is demand driven and that the efficiency of agricultural production systems can be drastically increased through judicious use of inputs and advances in biotechnology. During the past decade, for example, fossil-fuel input in Chinese agriculture rose 100-fold, and crop yields tripled (Lu et al., 1982). The comparatively low output of Indian agriculture can be attributed to low energy influx. India annually uses 142 kilograms (kg) of per capita coal equivalent compared with 4,871 kg in the United Kingdom and 10,410 kg in the United States (Bureau of the Census, 1983).

Given that the world as a whole does have the capacity to feed itself, what are the issues to be addressed and strategies to be adopted to achieve that goal? First, soil resources and population are unevenly distributed. Regions and countries with high demographic pressures are also characterized by low available land reserves, for example, South Asia, China, southeastern Nigeria, Rwanda, Burundi, the East African highlands, Central America, and the Caribbean. In some of those places, the daily per capita calorie intake is likely to remain below 2,500 at least through the year 2000 even with earnest efforts to improve agricultural production (Table C-2).

Second, even if technical know-how exists, socioeconomic, cultural, and political considerations are often overwhelming and do not readily permit the adoption of improved science-based technologies. The potential for increased fertilizers, pesticides, improved farm implements, and other innovations is limited due to nonavailability, high cost, or both. Often the major problems are poverty and lack of resources. Subsistence farmers will only use improved inputs if they are available at affordable prices.

Third, the overdependence on nonrenewable sources of energy is a global issue. All intensive systems of agricultural production are based on the use of fossil-fuel energy. In developed countries, such as Germany, the number of persons fed from 1 ha of cultivated land increased 5.6 times and the equivalent cereal yield increased 6.3 times between 1800 and 1978 (Mengel, 1990). This dramatic increase in agricultural production has been realized through the heavy use of fertilizers and other inputs. The United States invests about half of its fossil-energy input in agricultural production into supplying water (20 percent) and fertilizers (30 percent) (Pimentel, 1989). The annual amount of harvested nutrients in three major cereals (rice, wheat, and corn) is estimated at 40.1 million tons (t) of nitrogen, 15.32 million t of phosphorus, and 28.2 million t of K_2O. An equivalent

TABLE C-2 Food Availability: Calories Per Capita Per Day

Region	1983–1985	2000
Africa (sub-Saharan)	2,050	2,190
Near East/North Africa	2,980	3,100
Asia	2,380	2,610
Latin America	2,700	2,910
Low-income countries (excluding China)	2,130	2,350

SOURCE: Food and Agriculture Organization. 1989. The State of Food and Agriculture. Rome, Italy: Food and Agriculture Organization of the United Nations.

amount may be harvested in stover. The nutrients harvested must be replenished one way or another.

With the current level of world yields, the annual crop uptake is estimated at 85.5 kg of nitrogen per person (Andow and Davis, 1989), which will amount to a total of 530 billion t of nitrogen uptake by crops by the year 2000. Since inputs are an inevitable consequence of ever-increasing demand for agricultural production, several strategic issues must be resolved. Can nitrogen and other essential plant nutrients (for example, phosphorous, zinc, sulphur) be synthesized from the available reserves of fossil fuels? How can alternative sources of fertilizers or power be developed to meet the energy needs of developing countries? Are organic manures a viable source of nutrients required for agriculture in developing countries? It is estimated that only 2.5 percent of nitrogen in the manure is recoverable and usable with current technology (Pimentel, 1989). Moreover, the losses of nitrogen from organic manures by volatilization (30 to 90 percent) or leaching are major sources of water and atmospheric pollution.

In developing countries as a whole, only 4 percent of total commercial energy is used for agriculture, and merely 2.7 percent is in the form of fertilizer. Fertilizer use on arable land ranges from 4 to 50 kg/ha in most developing countries compared with 100 to 800 kg/ha in developed countries (Stout, 1989). Increasing use of fertilizer and other agricultural amendments is limited due to the restricted availability and high cost of nonrenewable sources of energy. Low cost and renewable hydroelectric power is available only in a few countries. Further, this premium form of energy is highly valuable and is preferably used for industrial purposes. About 70 percent of the world's nitrogen fertilizer is produced by using natural gas as the source of energy (Stout, 1989). Similar to population and land-resource availability, natural gas deposits are also unevenly distributed. Countries without natural gas have to import fertilizers.

Above all, the environmental issues of intensive agriculture cannot be ignored. In addition to the dangers of agricultural chemicals, the problem of deforestation in the tropics is a major environmental issue. Bringing new land under production through deforestation of tropical rain forests, as noted, has severe ecological, environmental, and sociopolitical implications. The actual extent of deforestation in the tropics is still the subject of debate, however (Myers, 1981). In addition to loss of biodiversity and potentially valuable genetic resources, rain forest conversion presumably contributes a large proportion of total global emissions of carbon dioxide (Houghton et al., 1987; Lashoff, 1988; Tirpak, 1988), although the exact values are not known. The type, amount, and rate of gaseous emission also depend on the method of deforestation—for example, slash and burn, chain-saw clearing, bulldozers, and chemical poisoning—and on the subsequent land uses (Lal, 1987a,b).

Then there is the problem of soil degradation. Currently, 5 to 7 million

ha of arable land (0.3 to 0.5 percent) is lost every year through soil degradation. The projected loss by the year 2000 is 10 million ha annually (0.7 percent of the currently cultivated area). A high proportion of this loss occurs
in ecologically sensitive regions of the tropics and subtropics, where marginal lands are being intensively cultivated.

Pollution of surface and groundwater by agricultural chemicals is another major environmental hazard. A high proportion of the fertilizer that is
applied is susceptible to volatilization, washed away in surface runoff or
eroded soil, or leached into the ground water. There is equal concern over
contamination of surface and ground waters by water soluble pesticides,
such as aldicarb, ethylene dibromide, and atrazine. Although pesticide use
is rapidly increasing in developing countries, drinking water supplies are
scarce, rarely treated, and seldom tested for contaminants.

The intensification of agriculture also involves other severe risks of environmental pollution. The so-called greenhouse effect is directly linked to
agricultural activities, and soil-related processes play a major role in the
emission of greenhouse gases. More organic carbon is contained in the
world's soil (in the form of soil organic matter) than in the world's biota or
atmosphere (Sedjo and Solomon, 1989; Stevenson, 1982). Intensive land
use for seasonal crop production may lead to depletion of soil organic matter and release of carbon into the atmosphere. Burning, a basic tool in
traditional agriculture, releases large quantities of greenhouse gases into the
atmosphere. In addition to burning and deforestation, other agricultural
practices that result in higher greenhouse emissions from tropical ecosystems include use of rice paddies (a major source of methane); intensive use
of marginal lands without inputs, which leads to mining and depletion of
soil organic matter; uncontrolled and excessive grazing with high stocking
rates; and indiscriminate use of chemical fertilizers.

In sum, the principal issues regarding soil research in the tropics with
relevance to agricultural sustainability are (a) food security related to the
perpetual deficit in some regions and widespread poverty and malnutrition
in others; (b) land scarcity and low carrying capacity of land; (c) soil degradation due to accelerated erosion, desertification, and salinization; (d) pollution and eutrophication of natural waters; (e) heavy reliance on nonrenewable fossil fuel for certain production technologies, and (f) possible greenhouse
effect due to deforestation, burning, and emission of radiatively active gases
into the atmosphere by soil-related processes.

SOIL-RELATED CONSTRAINTS TO
AGRICULTURAL PRODUCTION

Soil, the most basic of all resources, is finite on a global scale, nonrenewable in the human time frame, and extremely fragile and vulnerable to

misuse and mismanagement. This section describes soil-related constraints to intensive land use.

Soil Erosion

Accelerated erosion is a serious problem in several ecologically sensitive regions: the Himalayan-Tibetan ecosystem, the Andean region, the Caribbean, eastern Africa, and other densely populated regions with severe land shortage. Steeplands, which make up a large percentage of the total land area in these regions, are overexploited and grossly misused.

High erosion rates are observed throughout the tropics (Table C-3). In South and Southeast Asia, rivers draining the Himalayan region (for example, the Ganges, Mekong, Irrawdy, and Brahmaputra) have a high sediment load (Table C-4). In India, 150 million ha are subject to accelerated

TABLE C-3 Selected Erosion Rates in the Tropics

Region/Ecology	Criteria	Equivalent Field Erosion Rates (t/ha/yr)
Africa		
Cote d'Ivoire	Bare soil	138
Ethiopia	Sediment load	165
Ghana	Bare soil	100–313
Lesotho	Sediment load	180
Nigeria	Bare soil	230
Tanzania	Bare soil	38–93
Asia		
Bangladesh	50% slope	520
India	Cropland,	4–43
	gullies	33–80
Java, Indonesia	Imperata	345
Tropical America and the Caribbean		
Colombia	Cropland	21.5
El Salvador	Steeplands	130–260
Guatemala	Steeplands cultivated in maize	200–3,600
Northeast Brazil	Cropped land	115
Peru	Bare soil	148
Trinidad	10–20º slope, bare	490

SOURCES: R. Lal. 1986a. Conversion of tropical rain forest: Agronomic potential and ecological consequences. Adv. Agron. 39:173–264. Reprinted with permission by Springer-Verlag © 1986. R. Lal. 1986b. Soil surface management in the tropics for intensive land use and high and sustained production. Adv. Soil Sci. 5:1–138. Reprinted with permission by Springer-Verlag © 1986.

TABLE C-4   Sediment Yield from Some Tropical and Subtropical Catchments

Country	River	Sediment Yield ($t/km^2/year$)
China	Dali	16,300–25,600
Indonesia	Cilutung	12,000
Kenya	Perkerra	19,520
New Guinea	Ause	11,126

SOURCE: R. Lal. 1990c. Soil Erosion in the Tropics: Principles and Management. New York: McGraw-Hill. Reprinted with permission.

soil erosion (United Nations Environment Program, 1983). Siltation of reservoirs in northern India is about 200 percent more than was anticipated in their design (Table C-5; Dent, 1984). In Nepal, 63 percent of the Shivalik zone, 26 percent of the middle mountain zone, 48 percent of the transition zone, and 22 percent of the high Himalayas are subject to severe erosion. In Pakistan, the upper Indus basin is severely eroded. In China, about 46 million ha of the loess plateau are subject to severe erosion, which is raising the bed of the Yellow River by as much as 10 centimeters annually. Severe erosion is also occurring in the watersheds of the Yangtze, Huaihe, Pearl, Liaolie, and Songhua rivers (Dent, 1984). In South America, about 39 million ha or 8 percent of the Amazon basin are characterized by soils of high erodibility (Sanchez et al., 1982). In Africa, as much as 1 billion t of topsoil are lost from Ethiopian highlands each year (Brown, 1981), and the average annual rate of soil erosion from Madagascar is reported to be 25 to 40 t per ha (Finn, 1983). The Food and Agriculture Organization (1983)

TABLE C-5   Annual Rates of Siltation in Selected Reservoirs in India

Reservoir	Assumed Rate ($ha-m/100 km^2$)	Observed Rate ($ha-m/100 km^2$)	Year of Observation
Bhakra	4.29	6.00	1975
Ghod	3.61	15.51	1970
Mayurkashi	3.61	20.09	1975
Msiyhon	1.62	13.02	1971
Nizam Sagar	0.29	6.57	1967
Panchet	2.47	9.02	1974
Ramganga	4.29	17.30	1973

SOURCE: National Land Use Conservation Board. 1986. Review of Centrally-Sponsored Schemes of Soil Conservation in the Catchments of River Valley Projects. New Delhi: Government of India.

estimates that 87 percent of the Near East and Africa north of the equator are subject to accelerated erosion.

Wind erosion is equally severe in arid and semiarid regions (for example, the West African Sahel, western India, and Pakistan). In southern Tunisia, Floret and Le Floch (1973) and Le Houerou (1977a,b) observed that wind erosion rates of 10 millimeters of topsoil removed per year are common. Wind-blown dust from the Sahara causes air pollution and "sand rains" in the Caribbean (Rapp, 1974) and in northern Europe (Le Houerou, 1977a,b). It is estimated that between 25 and 37 million t of African soil are blown across the Atlantic Ocean annually (Prospero and Carlson, 1972). The global area subject to desertification is estimated to be 37.7 million square kilometers (km²)—16.6 million km² of the world's arid regions, 17.1 million km² of the semiarid regions, and 4.0 million km² of the subhumid regions (Mabbutt, 1978). The global loss to desertification is estimated at 6 million ha annually, and the rural population severely affected by desertification is about 135 million (United Nations Environment Program, 1984).

In several countries strong evidence exists of severe loss in soil productivity due to accelerated erosion. Instances of permanent soil productivity loss due to human-induced water erosion have been reported in several countries of Asia and Africa (Dregne, in press). Loss in agricultural production depends on soil properties, crops, management, and climate. In some shallow soils, the loss can be 50 percent or more (Lal, 1986b).

Structural Deterioration and Soil Compaction

An important process leading to soil degradation is the deterioration of the soil's structural properties and its ability to regulate water and air movement through the profile. Structural degradation, caused by a decline in soil organic matter and clay content and reduction in biotic activity, leads to crusting, compaction, reduced infiltration rate and low available water-holding capacity, increased soil detachability, and accelerated runoff and soil erosion. High erosion risk is a direct consequence of deterioration of soil structure.

Soil compaction is a more severe problem in soils of the tropics than hitherto anticipated. Soils with low-activity clays have slight or negligible swell/shrink capacity. Decline in soil organic matter content, degradation of soil structure, and excessive drying accompanied by high soil temperatures generally lead to consolidation and compression by mechanisms not well understood. Over and above these factors is the compactive effect of heavy machinery. It is estimated that some 90 percent of the soil surface may be traversed by tractor wheels during, for example, the primary tillage operations (Soane et al., 1981). A smearing action of the plow sole results in pore discontinuity that inhibits water movement and root development.

Characterization of soil compaction is another problem that is particularly severe for heterogeneous, gravelly soils. The usual criteria, such as bulk density, total porosity, and penetrometer resistance, are not the best indicators of the problem that plant roots experience. Perhaps the void ratio, the specific volume, air:water permeability, or pore-size distribution and continuity may be better indices of plant response than the bulk density. The available research information on these aspects of soils in the tropics is rather scanty. Critical soil bulk density values for root penetration and crop growth are not known for major soils of the tropics. Crops susceptible to soil compaction include maize (*Zea mays*), upland rice (*Oryza sativa*), sorghum (*Sorghum bicolor*), groundnut (*Arachis hypogea*), cassava (*Manihot esculenta*), yam (*Dioscorea rotundata*) and cowpea (*Vigna unguiculata*).

Decline in Soil Organic Matter Content

Rapid decline in the soil organic matter content of cultivated soils is a direct effect of continuously high temperatures throughout the year, low input agriculture, and soil erosion. Some studies have shown that the rate of mineralization of organic matter content in tropical soils may be four times greater than in temperate soils (Jenkins and Ayanaba, 1977). Consequently, cultivated soils in the tropics may have lower levels of organic matter than similar soils in temperate latitudes. Lal and Kang (1982) reported large differences in the organic carbon status of soils from various ecological regions of Nigeria: forest (1.3 ± 0.08 percent) > derived savannah (0.89 ± 0.071 percent) > Guinea savannah (0.7 ± 0.06 percent). The organic matter content of a soil and its susceptibility to erosion are intimately linked. Although a decrease in organic matter content increases the susceptibility of the soil to erosion, water erosion also preferentially removes soil colloids, including the humified organic matter fraction (Lal, 1976). Lal (1980) reported a linear decline in soil organic matter content with accumulative soil erosion:

Organic carbon (%) $= 1.79 - 0.002\,E$, $r = -0.71$,

where E is the annual accumulative soil erosion in tons per hectare. A decrease in organic matter content of the soil also increases its susceptibility to formation of surface crust, which further enhances the risk of soil erosion. Soil erosion is also increased by the reduction in biotic activity of soil fauna that occurs with a decrease in soil organic matter content.

In addition to decreases in structural stability, reductions in organic matter content have important implications in terms of plant-available water reserves in the soil. The favorable effects of organic matter content on the soil's water-retention capacity have been widely reported for soils of the tropics and subtropics (Lal, 1986b). In fact, organic matter content may have more beneficial effects on the available water-holding capacity than the clay content. Also important are the nutritional implications, for example, the effects on cation exchange capacity, acidification, and plant nutrients.

Fertility Depletion and Leaching

Continuous and intensive cropping with low or no off-farm input, necessitated by land hunger and poverty, cause fertility depletion and low yields. Many of the soils cultivated by shifting cultivators and subsistence farmers of the tropics and subtropics are subject to fertility depletion through decline in soil organic matter, reduction in nutrient reserves by crop removal, leaching, and acidification. Leaching and acidification are serious problems in soils of tropical climates with seasonally humid (alfisols) and humid moisture regimes (ultisols and oxisols). Substantial areas of acid tropical soils occur in Sumatra, Malaysia, the Congo basin, the Amazon basin, and in the *cerrados* and *llanos* of Brazil and Colombia. Nitrogen is most readily lost. The extent of loss can be as high as 60 kg/ha annually from cropped land and 300 kg/ha annually from uncropped land (Suarez de Castro and Rodriguez, 1958). In Queensland, Australia, Martin and Cox (1956) reported leaching, with losses of 27 kg of nitrogen/ha from vertisols, in subhumid environments.

Salt Imbalance

Salt-affected soils, totaling about 323 million ha are widely distributed throughout the arid and semiarid regions (Table C-6; Beek et al., 1980). The problem is particularly severe in irrigated regions of China, India, Pakistan, and the Middle East. Productivity of irrigated lands in these regions is severely jeopardized by salt imbalance in the root zone (Gupta and Abrol, 1990). Soil structure is adversely affected by the predominance of sodium and application of irrigation water of poor quality (Gupta and Abrol, 1990; Mathieu, 1982). Nonavailability of good-quality irrigation water is a severe constraint to expanding irrigated agriculture in arid regions.

Drought Stress

Only 2.5 percent of the world's water is freshwater, and only a fraction of that is available for agricultural purposes. Total annual global precipitation is estimated at $350 \times 10^3 \, km^3$, of which 78.6 percent ($275 \times 10^3 \, km^3$) falls over the oceans. Of the remainder, 64 percent evaporates, leaving merely $28 \times 10^3 \, km^3$ for surface runoff or groundwater (Hall, 1989). Scarcity of freshwater is a general problem, especially in countries with arid and semiarid climate (annual rainfall of less than 700 mm). In addition to the low total amount, rains in such regions are highly irregular and seasonal.

TABLE C-6 Global Distribution of Potential and Actual Areas of Salt-Affected Soils

Region	Area $(10^6$ ha)
Australia	84.7
Africa	69.5
Latin America	59.4
Near and Middle East	53.1
Europe	20.7
Asia and Far East	19.5
North America	16.0
World	322.9

SOURCE: K. J. Beek, W. A. Blokhuis, P. M. Driessen, N. Van Breemen, R. Brinkman, and L. J. Pons. 1980. Pp. 47–72 in Problem Soils: Their Reclamation and Management. ILRI Publication No. 27. Wageningen, Netherlands: International Institute for Land Reclamation and Improvement. Reprinted with permission.

Worldwide, only 2 percent of the cropland is fully or partly irrigated (Table C-7). Most of the irrigated land is in China, India, and Pakistan. In sub-Saharan Africa, only 3 percent of the cropland is irrigated. Further expansion of irrigated land is severely constrained. Indeed, the rate of increase in irrigation is decreasing in all regions of the world except sub-Saharan Africa, but the percentage increase observed there is misleading because the total irrigated area is rather low. In some countries, by contrast, agriculture is based almost entirely on irrigation (for example, 100 percent of the farmland in Egypt, 72 percent in Pakistan, and 67.4 percent in Japan). Despite intensive use of irrigation, irrigated land in the world constitutes only 0.04 ha per capita (Cervinka, 1989).

Although drought is a natural phenomenon in arid and semiarid climates, its effects and duration are accentuated by human-induced changes in the ecosystem. The effective use of rainfall is drastically reduced by anthropogenic soil degradation—compaction, crusting, erosion, nutrient imbalances, and so on.

RESEARCHABLE PRIORITIES

Improvement in subsistence farming can be realized through increasing production, sustaining the higher level of net output, and preserving the productive potential of natural resources through restorative measures of soil and crop management. The desired high net output must be achieved

TABLE C-7 Percentage of Harvested Land Under Different Water Regimes

Category	Worldwide	Sub-Saharan Africa	Near East/ North America
Low-rainfall rain-fed land[a]	8	17	14
Uncertain rainfall rain-fed land[b]	13	20	11
Good rainfall rain-fed land[c]	24	31	19
Problem lands, excessive rainfall[d]	22	26	16
Naturally flooded land[e]	11	3	11
Irrigated land[f]	22	3	29
Total	100	100	100

[a] 1–119 Growing days.
[b] 120–179 Growing days.
[c] 180–269 Growing days.
[d] >269 Growing days but soil quality may be marginal.
[e] Land under water for part of the year.
[f] Fully or partially irrigated.

SOURCE: Recalculated from Food and Agriculture Organization. 1990. Meeting of the Administrative Committee on Coordination and the Interagency Working Group on Water and Sustainable Agricultural Development, Food and Agriculture Organization of the United Nations, Rome, Italy.

with a minimum of soil degradation, however. The soil quality and its productive capacity must be preserved and improved by preventing soil erosion, promoting high biological activity of soil fauna, improving soil organic matter content, and replacing the nutrients harvested by crops and animals through chemical fertilizers and organic amendments, supported by effective nutrient-recycling mechanisms. The productive efficiency of a system must be evaluated in terms of its effect on the natural resources (for example, change in soil organic matter reserves, acidity-alkalinity balance, nutrient reserves, exchangeable cations, plant-available water capacity, and effective rooting depth). Suitable farming systems are those that enhance soil quality. Fertility mining and soil degrading, low-input systems must be stopped.

Insufficient water supply is the most important single factor governing agricultural production in arid and semiarid regions. Over and above water conservation in the root zone for rain-fed agriculture, supplemental irrigation of one form or another is necessary to reduce vulnerability of crops to adverse rainfall conditions and to increase crop yields. Except for South

Asia, China, and Egypt, irrigation potentials are not well defined (in sub-Saharan Africa, for example). Runoff in African river systems is low, and aquifers are meager compared with the vast rechargeable reserves in the Indus and Gangetic valleys. Management of large-scale irrigation schemes, similar to those of South Asia, is highly capital intensive. For these regions, small-scale irrigation schemes are appropriate and await development and expansion. Suitable combinations of soil management techniques (for water conservation and water harvesting/recycling) and supplementary irrigation are needed to enhance production and reduce risks of crop failure.

Simplified agricultural ecosystems are more productive, but they are often more susceptible to environmental stresses than natural ecosystems. The objective of sustainable management is to minimize the vulnerability of these systems to the degradative effects of accelerated erosion, rapid depletion of soil organic matter and nutrient reserves, and excessive buildup of unfavorable flora and fauna. Risks of instability or fragility created through the simplification of ecosystems are to be minimized through appropriate soil and crop management.

The extensive agricultural systems based on fertility-restorative measures involving shifting cultivation or bush fallow rotation are no longer economically viable or ecologically compatible. Sustainable systems are those that can produce economic returns on a continuous basis without causing large or long-term damage to the environment, and without being ethically or aesthetically unacceptable. Local, specific, and on-farm validation is needed in adapting the research information already known. There is also a need to create new knowledge through biotechnology and other modern innovations. Nonetheless, several proven subsystems or components are available and can be used as building blocks for formulating sustainable systems for a range of agroecosystems. A number of researchable priorities are described below.

Nutrient Management

Nutrient management is crucial to sustained production. Highly weathered oxisols/ultisols and alfisols, being inherently low in nutrient reserves, must have a regular and supplemental nutrient supply if they are to be intensively cultivated for increased food production. Intensive land use and high yields on soils of low inherent fertility can only be achieved by raising the nutrient levels through the use of inorganic fertilizers, organic amendments, and nutrient recycling. Nutrient enhancement for such soils is indispensable.

Although crop production can be increased by increasing fertilizer use, many small landholders and resource-poor farmers cannot afford expensive fertilizers. Policymakers and economic planners must develop long-range strategies to ensure a dependable and timely supply of chemical fertilizers at affordable prices. Alternative techniques must also be developed to avoid overdependence on synthetic fertilizers and other agricultural chemicals. The energy and economic costs of such a strategy are prohibitive for the small landholders of the tropics. Agronomic experiments must assess the appropriate combination of inorganic and organic fertilizers to minimize dependence on synthetic fertilizers and enhance soil structure and physical characteristics. Techniques must be developed to reduce the rate of application of inorganic fertilizers by minimizing losses and increasing the recycling of nutrients. In this regard, it is important to quantify losses by volatilization, leaching, and erosion in relation to conservation tillage, application by split doses, fertilizer placement, and slow-release formulations. Technological options for nutrient recycling must be researched for crop residue management and mulch farming, legume-based rotations, ley farming with different stocking rates and controlled grazing, and agroforestry systems, including alley cropping. Nutrient-recycling mechanisms and effects must be assessed for different soils (for example, highly weathered oxisols and ultisols, which predominantly contain aluminum [Al^{+3}] and manganese [Mn^{+3}] in the subsoil horizons and are devoid of basic cations). The effects of alley-cropping systems on crop yields should be evaluated in terms of competition for nutrients, water, and light (Lal, 1989, 1990a). Advantages in substituting biological nitrogen fixation for inorganic fertilizers must be quantified. And finally, careful evaluation is needed of the economics of growing nitrogen versus buying nitrogen, especially with regard to timely availability, land scarcity, efficiency of nitrogen from biological resources, and environmental effects.

Erosion Management

Erosion management is crucial to the sustainable management of soil resources. Several technological options are available for erosion management. A stronger data base and appropriate criteria are needed, however, to guide the choice of appropriate options, including due consideration of soil types, land form and terrain characteristics, rainfall regime and hydrology, cropping/farming system, and socioeconomic factors.

The pros and cons of measures to prevent versus control erosion should be carefully assessed. Preventive measures are those that enhance soil structure, decrease raindrop impact, improve infiltration capacity, and decrease runoff rate and amount. Use of these techniques is based on thorough knowledge of soil and crop management (for example, mulch farming through cover crops and planed fallows, multiple cropping, multistory canopy including agroforestry, and conservation tillage).

Vegetative hedges are important tools in minimizing risks of soil erosion (Table C-8). Although general principles are known, there is a need to validate and adapt these practices under locale-specific conditions for major

TABLE C-8 Effects of Contour Hedges of *Leucaena* and Vetiver on Runoff and Soil Erosion from Shallow Soil Planted to Pearl Millet and Deep Soil Planted to Sorghum in Central India

Treatment	Grain Yield (t/ha)	Runoff (%)	Soil Erosion (t/ha)
Pearl millet on a shallow soil			
Across-the-slope sowing	1.5	17.7	11.5
Contour cultivation along *Leucaena* keyline	1.7	11.8	6.2
Contour cultivation along vetiver keyline	2.0	9.0	3.3
Sorghum on a deep soil			
Across-the-slope sowing	3.4	21.5	18.4
Contour cultivation along *Leucaena* keyline	3.7	18.1	9.4
Contour cultivation along vetiver keyline	3.9	3.7	4.3
Cultivation along with graded bunds	3.5	17.3	14.2

SOURCE: Manoli Watershed Development Project. 1990. Pungabrao Krishi Vidyepeeth: A report on research highlights of technical programme. Annual Report, Manoli Watershed Development Project. Photocopy.

crops, cropping systems, soils, and ecological regions of the tropics. The adaptability of local tree species as vegetative hedges for erosion control and other uses should be evaluated. The adaptability of tillage systems to erosion control must be judged in terms of socioeconomic and cultural factors, availability and maintenance of implements, cost and availability of herbicides, and efficiency of soil and water conservation. The ecological limits of different conservation tillage methods must also be established.

Residue Management

A regular and sizable addition of organic material to soil is essential to maintain favorable organic matter content and to stimulate biotic activity of soil fauna, including earthworms and termites. Structural collapse of soils with predominantly low-activity clays can be avoided by maintaining high organic matter content and by enhancing the activity of soil fauna. Crop residue mulch is an important ingredient of any improved farming-cropping system. Although the beneficial effects of mulching are widely recognized, procuring a mulch material in sufficient quantity is a serious practical problem. Research on management of crop residue as a source of mulch must be closely linked with cropping systems, tillage methods, and planted fallows. Appropriate cultural practices must be developed and validated to ensure an adequate amount of residue mulch for soil protection and fertility enhancement. Live mulch, alley cropping, ley farming, planted fallows, and the use of industrial by-products are some of the cultural practices for procuring mulch that need to be validated. Their suitability depends on locale-specific biophysical and socioeconomic environments.

Crop Management

It is widely recognized that continuous ground cover is necessary to provide a buffer against sudden fluctuations in micro- and meso-climate and to prevent the degradative effects of raindrop impact or high-velocity winds. Ensuring protective ground cover requires research information on appropriate time of planting, optimum seed rate, improved cultivars and cropping systems, fertilizer use, pest controls, and other important aspects of crop management. The benefits of timely planting must be assessed against uncertain rains, unfavorable soil temperature regime, pest infestation, and unfavorable markets. Planted fallows, using both legume and grass covers, must be evaluated for their restorative effect on soil fertility and soil physical properties in comparison with natural fallows. How long does it take improved soil organic matter content to affect soil structure favorably?

Fallow Management

When crop residue mulch is inadequate, practical means must be developed to procure mulch through incorporation of an appropriate cover crop or planted fallow in the rotation. In addition to their capacity to supply residue mulch, planted fallows must also be evaluated for their usefulness in restoring physical and nutritional properties in comparison with long bush fallows. Information is needed on appropriate species of cover crops and methods for their management. Other researchable topics in fallow management include the timing and methods of suppression of the fallow crop, the timing and methods of sowing the food crop, methods of weed control and competition reduction between the fallow and food crop, and the timing and amount of nutrient released by the fallow crop.

Hedge-Row Management and Agroforestry

Agroforestry, as a special case of mixed cropping, involves growing deep-rooted perennial leguminous shrubs and trees in association with food crop annuals or livestock. This practice supposedly minimizes the soil-degradation risks of intensive use of arable land (King, 1979; Vergara,

1982). Perennial crops that fruit annually, such as banana, may be more suitable for intercropping with annuals than plantation crops. Shade-tolerant staples, such as cocoyam, can be grown in association with plantation crops (for example, cocoa), especially in the early stages of plantation establishment or along the outer margins. Research must ascertain which species of trees and woody perennials are most appropriate and can be profitably grown in association with annuals or animals. More information is also needed on management systems for perennials and annuals that maximize their benefits and reduce competition. The amount of nutrients contributed by perennials depends on soil and crops; research must determine those amounts. In addition to nutrients, more needs to be learned about the water requirements of perennials and annuals. Allelopathic effects, if any, must be carefully assessed.

Water Management

Water management is critical in alleviating the adverse effects of recurring drought on crop and animal productivity. Lack of water during the growing season, like the lack of nutrients, is a major constraint in arid and semiarid climates. Agricultural productivity in several regions of the tropics and subtropics primarily depends on the amount, distribution, and reliability of rainfall. Efficient use of rainfall is crucial to sustainable productivity in rain-fed agriculture. An understanding of the rainfall characteristics of a region—the probability of the occurrence of a certain amount at a desired frequency, or the onset of assured rains at a given time—is an important requisite for developing sustainable systems of crop and soil management in that region. Controlling and managing runoff is a key management objective. In addition to conserving rainwater in the root zone by decreasing losses due to runoff and evaporation, means of supplementing irrigation must be explored to decrease the sensitivity of production to climate.

Irrigation, a capital-intensive technology, has not been fully exploited in several regions. It is estimated, for example, that only 2 percent of the irrigable land in Africa is irrigated. In addition to the development of feasible large-scale irrigation schemes, high priority should be given to small-scale, labor-intensive schemes. Small-scale irrigation schemes may be more appropriate for resource-poor farmers and in regions where large rechargeable aquifers do not exist. Replacement of traditional devices (for example, shadoof, Persian wheel driven by animal) by diesel, electric, or wind-driven pump may improve the efficiency and increase the cropped area under irrigation. The technical and social issues related to water delivery, water allocation, and water-use efficiency must also be addressed, however. Each of the alternatives involves a different set of problems and possible remedies, and their socioeconomic and political dimensions must be taken into account.

Watershed Management

Sustainable management of soil and water resources is based on judicious and scientific management of all landscape units within a watershed. Widespread and severe problems of accelerated erosion and sedimentation, perpetually devastating floods, land degradation beyond the point of no return, eutrophication of water, and environmental pollution in general are traceable to poor planning and mismanagement of landscape units within watersheds. Scientific criteria for the choice of appropriate land uses, exploitation of water resources for irrigation and domestic purposes, and the development of infrastructure (including access roads) must be developed. Scientific use of a watershed for sustainable land and water development is more easily described than achieved, however. The problem is caused in large part by private ownership of small landholdings; farm boundaries cut across landscape units and natural waterways. The problem is aggravated by dubious land-tenure systems and ownership rights. Legislation, policies, and incentives are needed to foster cooperation among farmers and promote ecologically compatible development of natural resources.

SYSTEMS APPROACH

Although general principles may be the same, technological packages (systems) for sustainable management of soil and water resources are site specific and depend on farming-cropping systems, farm size, availability of essential inputs, and socioeconomic factors. Locale-specific and on-farm synthesis of packages is needed on the basis of the components and subsystems described above. The agronomic productivity, economic profitability, and ecological compatibility of such packages must be assessed through appropriate research. Systems research is preferably conducted on "benchmark" soils or "ecological regions." In that way, the agroeconomic productivity of different production systems can be related to soil and climatic characteristics. This approach will facilitate transfer of technology to similar soils and environmental conditions elsewhere. Systems research necessitates a pan-disciplinary approach involving scientists with expertise in soil science, hydrology, climatology, agricultural mechanization, agronomy, plant improvement, pest management, economics, sociology, and anthropology. Results obtained from field experimentation can be validated against predictive models. The latter may be biophysical models, economic-productivity models based on linear programming, or statistical models based on systems analysis of empirical data.

The alternative agronomic approach involves field experimentation and gradual, step-by-step improvement of the system through substitution of the component that is the major constraint to crop and animal production. The aim of on-farm research is to identify the major constraint and alleviate its effect by devising technological options. The agronomic approach is a long-term strategy aimed at transforming low-input subsistence farming into science-based agriculture. Researchable priorities in this approach involve assessment of the components or subsystems under on-farm conditions and with the active involvement of farmers. In addition, specific research priority should be given to soil and crop management practices that increase the efficiency of water and fertilizer use and restore eroded and degraded lands.

CONCLUSIONS

Soil research must be mission oriented. Its objective is the alleviation of production-related constraints in intensive agriculture. In that context, sustainable management of soil and water resources implies meeting current needs without jeopardizing future potential. Thus, sustainability of soil and water resources must be judged with tangible criteria—soil and water conservation, productivity, restoration of degraded soil, reduction in off-farm inputs for the same level of production and profitability, increase in labor productivity, and so forth. Major considerations in terms of research and development priorities are outlined below.

• Farming systems and technologies that enable people to live in comfort and in symbiosis with nature must be developed. Many misconceptions still persist about the actual potential and constraints of tropical ecosystems. Although considerable progress has been made in the recent past toward replacing myths with facts, the reasons for the lack of sustainable-yield farming systems in tropical ecosystems are not fully understood. The objective of research and development is to achieve high and sustainable yields, but with low inputs and with reduced damage to soil and environments. The goal is to achieve optimum sustainable yields with modest inputs, rather than maximum yields based on high capital and energy inputs.

• The first research priority should be to determine why the research information already available is inadequate. Although the innovative concepts and subsystems proposed in the literature are technically sound, their economic evaluation and social acceptability must be assessed for varying socioeconomic environments. These concepts should be evaluated as integral parts of the overall system rather than as individual components of the improved technology. Improved soil management is sustainable only within economically improved farming systems.

• The promising innovations and improved subsystems already developed to alleviate specific biophysical constraints related to soil and environments should be integrated into farming systems as specific case studies. Methodologies involving linear programming and systems analysis should be standardized to facilitate establishment of blueprints of farming systems for locale-specific situations on the basis of subsystems and component technology already developed.

• Baseline data are needed on the soil, climate, water, and vegetation resources of the tropics, and their potentials and constraints. Resource data banks must be established for major agroecological zones, and their productive potential of the zones must be defined on the basis of conceptual models. Land-use planning is the key to sustainable development of land and water resources, and it requires systematic surveys of soil, hydrology, vegetation, and terrain at a practical scale (<1:50,000). Few, if any, developing countries have programs for detailed and systematic evaluations of natural resources at this scale. The emphasis on these surveys is not for soil classification and mapping per se, but for assessing the potential of, and constraints on, natural resources and for developing management options for sustained production without causing degradation of fragile land resources. The data bank established on natural resources can be used to provide site-specific information for choosing technological options through the use of geographic information systems, global positioning systems, and digital elevation models.

• Rapid research progress has to be made by the year 2000, when the demands made on soil resources will be greater than ever before. The low-input systems now being recommended are already obsolete. Now is the time for researchers to provide the components or subsystems of medium-to high-input technology. Specifically, research for the twenty-first century must provide, as a top priority, basic data from well-designed, long-term field experiments on soil management that involve technologies with different levels of inputs. These experiments must address the following scientific concerns:

1. Restoration of eroded and degraded lands. This deserves high priority, particularly to reduce the need to clear and develop new land. Methods should be developed to restore and rehabilitate eroded and degraded lands. Land-evaluation criteria should indicate the time when soil should be taken out of production and put under a restorative and ameliorative phase. Knowing the critical limits of soil properties for different levels of management is crucial in this endeavor.

2. Soil compaction. The problem of soil compaction will become severe with intensive land use and increasing mechanization. There is a need to develop routine methods for characterizing soil compaction. No standardized procedures are available for minimizing soil compaction by motorized farm equipment or for restoring production on soils hitherto compacted.

The problem should be addressed from various aspects (for example, machinery, rotations, and farming systems). Traffic-induced soil compaction warrants an interdisciplinary research approach by soil and crop scientists working together with soil engineers and machinery designers.

3. Soil erosion. Additional basic data are needed on soil erosion and its control and on predicting water runoff and soil loss under different land uses and in different farming systems. In addition to basic factors affecting erosivity and erodibility, the numerical limits of "soil loss tolerance" should be established. This information is important for long-range planning for different land uses. Conceptual and empirical models relating crop yield and soil loss to different levels of management are needed for assessing the economic consequences of accelerated soil erosion. Unless the relationship between erosion and productivity is developed, it will be difficult to plan development strategies and choose soil management methods.

4. Soil management. Management of soil structure is still a mystery. Why does the structure of most soils of the tropics deteriorate so rapidly, and how can it be prevented? Development of crust and surface seal is a major problem in soils containing predominantly low-activity clays, including those in the semiarid tropics. Techniques of soil surface management should address this problem.

5. Increasing output. Increasing agricultural output per unit of input will remain a challenge for generations to come. The inputs may be natural resources or soil amendments. In the semiarid and arid tropics, increasing output per unit of water is the basic ingredient of a successful technology. In humid regions with highly leached soils, limited plant nutrients (for example, calcium, nitrogen, and phosphorus) are the major consideration in developing improved farming systems. Also to be studied are soil-water-fertility interactions for different agroclimatic regions, including leaching patterns and salt and water balance.

6. Tillage systems. The development of appropriate tillage systems for different-sized farms and for a wide range of soils, crops, and climatic environments is an important research consideration. Given the merits of no-till systems in controlling erosion and conserving water, their potential should be exploited. The ecological limits to the application of these no-till techniques can be greatly extended by improving the agronomic practices associated with their implementation. Tillage systems should be geared toward alleviating specific soil-related constraints to crop production—soil temperature, soil and water conservation, soil compaction, maintenance of soil structure, and soil organic matter content.

7. Soil dynamics. The evolution of the physical, chemical, and biological properties of soil should be studied in representative farming systems and under varied land uses in order to establish the cause-effect relationship

between land use and soil properties. The agronomic output should always be assessed in terms of the impact of land use on soil properties.

8. Soil constraints. The soils of humid regions have special toxicity constraints. Research is needed on how to adapt crops to these nutrient constraints. In arid and semiarid environments, on the other hand, salt accumulation in the surface soil horizon is a special constraint. Basic studies of salt and water balance in different farming systems and in different ecological regions should provide the basis of management systems to overcome these problems.

9. Nutrient recycling. To decrease inputs of chemical fertilizer, priority should be given to research on agroforestry techniques, planted fallows, and other nutrient-recycling systems, such as use of organic wastes and farm by-products.

10. Irrigation. There is a need to develop irrigation potential fully, especially in sub-Saharan Africa. Expansion of irrigable cropped area warrants high priority. Both the technical and social issues related to water delivery and water allocation must be addressed. Each of these involves a different set of problems and possible remedies, and each has socioeconomic and political dimensions that must be taken into account.

REFERENCES

Andow, D. A., and D. P. Davis. 1989. Agricultural chemicals: Food and environment. Pp. 192–235 in Food and Natural Resources, D. Pimentel and C. W. Hall, eds. San Diego, Calif.: Academic Press.

Beek, K. J., W. A. Blokhuis, P. M. Driessen, N. Van Breemen, R. Brinkman, and L. J. Pons. 1980. Pp. 47–72 in Problem Soils: Their Reclamation and Management. ILRI Publication No. 27. Wageningen, Netherlands: International Institute for Land Reclamation and Improvement.

Brown, L. R. 1981. World population growth, soil erosion, and food security. Science 214:995–1002.

Bureau of the Census. 1983. Statistical Abstract of the United States. 1983. 104th ed. Washington, D.C.: U.S. Government Printing Office.

Buringh. P. 1981. An Assessment of Losses and Degradation of Productive Agricultural Land in the World. FAO Workshop on Group Soils Policy. Rome, Italy: Food and Agriculture Organization of the United Nations.

Cervinka, V. 1989. Water use in agriculture. Pp. 142–163 in Food and Natural Resources, D. Pimentel and C. W. Hall, eds. San Diego, Calif.: Academic Press.

Dent, F. J. 1984. Land degradation: Present status, training and education needs in Asia and the Pacific. UNEP Investigations on Environmental Education and Training in Asia and the Pacific. FAO Regional Office. Bangkok, Thailand: Food and Agriculture Organization of the United Nations.

Dregne, H. E. 1990. Erosion and soil productivity in Africa. J. Soil Water Conserv. 45:431–436.

Dregne, H. E. In press. Erosion and soil productivity in Asia. J. Soil Water Conserv. 46.

Dudal, R. 1982. Land degradation in a world perspective. J. Soil Water Conserv. 37:245–247.

Food and Agriculture Organization and United Nations Environment Program. 1983. Guidelines for the Control of Soil Degradation. Rome, Italy: Food and Agriculture Organization of the United Nations.

Food and Agriculture Organization (FAO). 1986. Irrigation in Africa south of the Sahara. FAO Investment Centre Technical Paper 5. Rome, Italy: Food and Agriculture Organization of the United Nations.

Finn, D. 1983. Land use and abuse in the East Africa region. Ambio 12:296–301.

Floret, C., and E. Le Floch. 1973. Production, sensibilite et evolution dela vegetation et de milieu en Tunisie presaharienne. CEPE Document No. 71. Montpellier, France: Centre d'etudes phytosociologiques et ecologiques Louis Emberger.

Gupta, R. K., and I. P. Abrol. 1990. Salt-affected soils: their reclamation and management for crop production. Adv. Soil Sci. 11:223–288.

Hall, C. W. 1989. Mechanization and food availability. Pp. 262–275 in Food and Natural Resources, D. Pimentel and C. W. Hall, eds. San Diego, Calif.: Academic Press.

Herdt, R. 1988. Increasing crop yields in developing countries. Paper prepared for meeting of the American Agricultural Economic Association, The Rockefeller Foundation, New York.

Houghton, R. A., R. D. Boone, J. R. Hobbie, J. E. Melillo, C. A. Palm, B. J. Peterson, G. R. Shaver, G. M. Woodwell, B. Moore, D. L. Skole, and N. Myers. 1987. The flux of carbon from terrestrial ecosystem to the atmosphere in 1980 due to change in land use: Geographical distribution of the global flux. Tellus 398:122–139.

Hudson, W. J. 1989. Population, food, and the economy of nations. Pp. 275–301 in Food and Natural Resources, D. Pimentel and C. W. Hall, eds. San Diego, Calif.: Academic Press

Jenkins, D. S., and A. Ayanaba. 1977. Decomposition of C^{14} labeled plant material under tropical conditions. Soil Sci. Soc. Amer. J. 41:912–915.

King, K. F. S. 1979. Agroforestry and the utilization of fragile ecosystems. For. Ecol. Man. 2:161–168.

Lal, R. 1976. Soil Erosion Problems on an Alfisol in Western Nigeria and Their Control. IITA Monograph 1. Ibadan, Nigeria: International Institute of Tropical Agriculture.

Lal, R. 1980. Soil erosion problems on Alfisols in Western Nigeria. Effects of erosion on experimental plots. Goederma 25:215–230.

Lal, R. 1984. Soil erosion from tropical arable lands and its control. Adv. Agron. 37:183–248.

Lal, R. 1986a. Conversion of tropical rain forest: Agronomic potential and ecological consequences. Adv. Agron. 39:173–264.

Lal, R. 1986b. Soil surface management in the tropics for intensive land use and high and sustained production. Adv. Soil Sci. 5:1–138.

Lal, R. 1987a. Conversion of tropical rain forest: Agronomic potential and ecological consequences. Adv. Agron. 39:173–264.

Lal, R. 1987b. Managing soils of sub-Saharan Africa. Science 236:1069–1076.

Lal, R. 1990a. Agroforestry systems and soil and surface management of a tropical alfisol. I-V. Agroforestry Systems 8:1–6, 7–29, 97–111, 113–132, 197–215, 217–238, 239–242.

Lal, R. 1990b. Myths and scientific realities of agroforestry as a strategy for sustainable management for soils in the tropics. Adv. Soil Sci. 15: 91–137.

Lal, R. 1990c. Soil Erosion in the Tropics: Principles and Management. New York: McGraw-Hill.

Lal, R., and B. T. Kang. 1982. Management of organic matter in soils of the tropics and sub-tropics. Pp. 152–178 in Proceedings of the Twelfth Congress of ISSS. Wageningen, The Netherlands: International Society of Soil Scientists.

Lashoff, D. 1988. Global climate scenarios related to agriculture. Prepared for Workshop on Agriculture and Climate Change, March 1988, U.S. Environmental Protection Agency, Washington, D.C.

Le Houerou, H. N. 1977a. The nature and causes of desertization. Pp. 17–38. in Desertification, M. H. Glantz, ed. Boulder, Colo.: Westview Press.

Le Houerou, H. N. 1977b. The scapegoat. Ceres 10:14–18.

Lu, M., J. Ysheng, and S. Chenyueng. 1982. Typical analysis of rural energy consumption in China. Nongya Jingji Luenchung 4:216–223.

Mabbutt, J. A. 1978. The impact of desertification as revealed by mapping. Environ. Conserv. 5:45–56.

Manoli Watershed Development Project. 1990. Pungabrao Krishi Vidyepeeth: A report on research highlights of technical programme. Annual Report, Manoli Watershed Development Project. Photocopy.

Martin, A. E., and J. E. Cox. 1956. N studies on black soils from Darling Downs, Qld., Australia. J. Agric. Res. 7:169–183.

Mathieu, C. 1982. Effects of irrigation on the structure of heavy clay soils in north-east Morocco. Soil Tillage Res. 2:311–329.

Mengel, K. 1990. Impacts of intensive plant nutrient management on crop production and environment. Pp. in 42–52 in Transactions of the Fourteenth Congress of ISSS, Plenary Papers. Wageningen, The Netherlands: International Society of Soil Scientists.

Myers, N. 1981. The hamburger connection: How Central America's forests become North America's hamburgers. Ambio 10:3–8.

National Land Use Conservation Board. 1986. Review of Centrally-Sponsored Schemes of Soil Conservation in the Catchments of River Valley Projects. New Delhi: Government of India.

Pimentel, D. 1989. Ecological Systems, natural resources, and food supplies. Pp. 2–31 in Food and Natural Resources, D. Pimentel and C. W. Hall, eds. San Diego, Calif.: Academic Press.

Propero, J. and T. Carlson. 1972. Vertical and areal distribution of Saharan dust over the western equatorial North Atlantic Ocean. J. Geophys. Res. 77:5255–5265.

Purnell, M. F. 1986. Applications of the FAO framework of land evaluation for conservation and land use planning in sloped areas; potential and constraints. Pp. 17–31 in Land Evaluation for Land Use Planning and Conservation in Sloping Areas, W. Fiderus, ed. Publication 40 Wageningen, Netherlands: International Institute for Land Reclamation and Improvement.

Rapp, A. 1974. A review of desertization in Africa. In Water, Vegetation, Man. Secretariat for International Ecology, Stockholm, Sweden.

Sanchez, P. A., D. E. Bandy, J. H. Villachica, and J. J. Nicholaides. 1982. Amazon basin soils: Management for continuous crop production. Science 216:821–827.

Sedjo, R. A., and A. M. Solomon. 1989. Climate and Forests. In Greenhouse Warming: Abatement and Adaption, N. J. Rosenberg et al., eds. Washington, D.C.: Resources for the Future.

Soane, B. D., P. S. Backwell, J. W. Dickson, and D. J. Painter. 1981. Comparison by agricultural vehicles: A review. I. Soil and Water Characteristics. Soil Tillage Res. 1:207–238.

Stevenson, F. J. 1982. Humus Chemistry: Genesis, Composition, Reaction. New York: John Wiley & Sons.

Stout, B. A. 1989. Handbook of Energy for World Agriculture. New York: Elsevier.

Suarez de Castro, G. K., and G. A. Rodriguez. 1958. Movimiento de agua en el suelo (estudio en Pisimentros nonliticos). Bol Tec. Fed. Nac. Cafeteros Colomb. No. 2.

Tirpak, D. 1988. Links between agriculture and climate change. Prepared for Workshop on Agriculture and Climate Change, March 1988, U.S. Environmental Protection Agency, Washington, D.C.

United Nations Environment Program. 1984. General assessment of progress in the implementation of the plan of action to combat desertification, 1978–84. Governing Council, 12th Session, Nairobi, Kenya. Photocopy.

Vergara, N. T. 1982. New Directions in Agroforestry: The Potential of Tropical Legume Trees. Honolulu, Hawaii: Environmental and Policy Institute.

APPENDIX D

The Agroecosystems

The proposed Sustainable Agriculture and Resource Management (SANREM) program is distinguished from its Collaborative Research Support Program (CRSP) predecessors by its focus on the sustainability of agroecological systems. Previously established CRSPs, and international agricultural research efforts in general, have focused on the development of technologies to increase the production of particular commodities. The commodity focus has enabled researchers to build interdisciplinary teams and methodologies, to strengthen institutional structures around the world, and to bring local experience and needs to the attention of the global research community. Not incidentally, commodity-centered programs have also yielded important insights into a variety of agronomic considerations—germplasm conservation, nitrogen fixation, rotational effects, and pest population dynamics, to cite only a few examples—that are critical to sustainable agriculture in their particular agroecosystems. The commodity focus has also enabled researchers to begin to define the social and economic issues associated with those particular commodities and the regions in which they are grown. The SANREM program must incorporate and build on the substantial record of achievement that those research programs have compiled.

The emergence of sustainability as an organizing concept and as a research objective is in itself evidence that, however effective in its special

This appendix is based on conclusions and research priorities identified by participants at the National Research Council's Forum on Sustainable Agriculture and Natural Resource Management, held in Washington, D.C., on November 13–16, 1990. Working groups discussed priorities in each of the four agroecosystems described below, and a fifth group addressed priorities across agroecosystems.

applications, commodity-oriented research is limited in its ability to embrace all the factors that influence the long-term health, productivity, and stability of the agroecological system as a whole. In focusing scientific attention on the productivity of a given plant crop, even broadly conceived research tends to neglect other crops, the full range of environmental influences on the crop, the environmental impact of the crop, the social and economic causes and effects of cropping systems, and the role of the crop in achieving a balanced and equitable system of land use. These factors also act as feedback mechanisms; they can eventually affect, both positively and negatively, the crops under scrutiny. The systems approach works to harness this understanding for the long-term well-being of the crop, people, and the system as a whole.

Agroecosystem research recognizes that the delineation of system boundaries implies a certain degree of flexibility. Boundaries must be defined clearly enough to allow for rigorous study, and loosely enough to take into account less immediate but still relevant factors affecting the system. In this sense, boundaries can vary depending on the scope of the hypothesis being tested. An investigation of the effects of intercropping on soil microbial activity in a single field, for instance, demands a different, though no less legitimate, scale of research than does a landscape-level investigation of the effects of cropping patterns on water quality. The agroecosystem approach enables researchers not only to adjust their focus to the scale appropriate for the hypothesis at issue, but to integrate hypotheses so that they may illuminate one another and the working of the system as a whole. The agroecosystem, in short, will be a fundamental concept in SANREM research, not only as an object of study, but as a *way* to study. It will serve as a tool to organize ideas, hypotheses, methods, and results, and ultimately to gain insight into principles of sustainability.

The resource base and the human population that relies on it are at greatest risk in several primary global agroecosystems: the humid tropics, semiarid range and savannah, hill lands, and input-intensive agroecosystems. In establishing and building the SANREM program, researchers will invariably focus on these systems. They offer the greatest potential for results that (a) are critical to environmental well-being, (b) can be broadly applied, (c) are relevant to great numbers of farmers, and (d) can help to define more precisely the characteristics of sustainability.

The above classification of primary global agroecosystems is not homogeneous. The first two systems are defined primarily by climatic and vegetational factors, the third by overriding topographical characteristics (slope, aspect, and elevation), and the last by input levels. Broad as they are, these categories necessarily obscure the incalculable diversity of local conditions within each—their ecological circumstances, characteristic biological diversity, historical land-use patterns, and cultural contexts. Moreover, certain sites legitimately belong to more than one category. Other systems, less extensive but nonetheless important in terms of human welfare, biodiversity, and other aspects of sustainability, may fit none of these categories. Such unavoidable shortcomings aside, this classification allows the broadest and most effective identification of problems, possibilities, gaps, and commonalities in the complex undertaking of research on sustainable agriculture and natural resource management.

HUMID TROPICS AGROECOSYSTEMS

Humid tropic agroecosystems are located in tropical regions where there is no more than a 3-month dry season and temperature is not a limiting factor for plant growth. The native vegetation in these areas is tropical rain forest. Rain forests once covered some 1.6 billion hectares (ha), principally in the low latitudes of Central America, South America, Africa, and Southeast Asia; smaller expanses existed along the eastern coasts of Madagascar, South America, and Australia. In recent decades, all of these areas have undergone rapid conversion. Approximately half of the former rain forest has been cleared for timber, fuelwood, farming, plantation agriculture, and cattle ranching. The remaining rain forests are concentrated in three large swaths: the Amazon basin in South America, the Congo basin in west central Africa, and the islands of the Malay Archipelago between Southeast Asia and Australia.

Indigenous agricultural techniques have evolved to fit the demands of the rain forest environment. Shifting cultivation enabled small populations of dispersed farmers to raise crops by clearing small plots in the forest, burning the slash for nutrients, raising a series of diverse crops in succession over a period of several years, and then allowing the plot to lie fallow to regain its forest cover and its vegetation-captured fertility. As population pressure has increased, however, shifting cultivation has become more prevalent and its time sequence more compressed. Under such conditions, clearing becomes more frequent and fallow periods decrease or disappear altogether. Large-scale logging, mining, plantation agriculture, and livestock ranching bring more settlers and more intensified land uses to the rain forests. These trends have placed ever greater pressures on the relatively poor tropical soils. Once cleared, the soils are easily eroded, their residual nutrients leached, and their role in the hydrological cycle disrupted. Soil acidity and nutrient deficiency are common chemical constraints to crop production in the humid tropics under any circumstances, and addressing those problems is key to the development of sustainable methods that can ease the pressure on the remaining rain forests, restore already degraded forestland through a variety of agroforestry strategies, and allow for more efficient and intensive cropping of developed land.

In any discussion of the future of the humid tropics, sustainable agriculture must be linked to the causes and consequences of deforestation. The forests of the humid tropics are currently being cleared at a rate that exceeds 10,000 square kilometers (km^2) per year. The rate of conversion has roughly doubled since the 1970s, when alarms about deforestation were first widely sounded. The destruction of tropical forests is of concern for three main reasons. First, the destruction of tropical forests currently releases between 25 and 30 percent of annual atmospheric carbon monoxide, carbon dioxide, nitrous oxide, and methane and less dramatic but still important quantities of other greenhouse gases. Although the industrial nations bear most of the responsibility for aggregate atmospheric carbon additions, the destruction of forests is particularly significant because important carbon sinks—tropical forest biomass and soils—are now becoming carbon sources. The models of greenhouse dynamics remain contentious, but they suggest that some of the areas of greatest agricultural production in the United States, especially California, Florida, and parts of the Midwest, are vulnerable to climate modifications. Deforestation also influences carbon emissions through its effects on local microclimates. As surface temperatures increase after conversion (often by more that 10 degrees centigrade), the breakdown of soil organic matter doubles. The consequent release of carbon from soils is orders of magnitude greater than that released by biomass burning. The long-term global changes and the potential dislocations they imply should make the control of deforestation an urgent priority.

The loss of biodiversity is a second consequence of current shifts in land use in the humid tropics. The rain forests, which now cover 7 percent of the earth's surface, are believed to contain at least 50 percent, and perhaps as much as 75 percent, of the total species diversity on earth. Deforestation is consequently bringing about the greatest destruction of the earth's organisms since the Cretaceous extinctions. There are strong ethical and economic reasons to avoid this annihilation of species. The potential economic returns from medicines, latexes, resins, and fibers are important, but they are eclipsed by the importance of wild genera of primary food and industrial crops. The loss of diversity of domesticated varieties has also accelerated as local farmers move off the land or adopt new varieties. The loss of biodiversity increases the vulnerability of both industrial and subsistence agriculture, and it narrows the base for commercial and subsistence plant breeders in developed and developing countries. The structure of U.S. agriculture will ultimately be affected, not only in areas that produce tropical crops (Hawaii, Puerto Rico, Florida, and California), but in virtually all regions in which major food and industrial crops are grown. The diminishment of floral and faunal diversity is cause enough for concern, but the displacement of forest peoples implies also an incalculable depletion of cultural diversity and indigenous knowledge.

Finally, the degradation of soil resources—through erosion, destruction of soil fertility, and loss of lands through urban and industrial encroachment—is advanced in the humid tropical areas. Although it is not the only factor leading to land abandonment and the poor performance of short-cycle tropical agriculture, soil degradation is a major contributing factor.

The above concerns make the careful management of tropical vegetation, soils, and water an urgent priority. Many current production systems, ranging from some kinds of shifting cultivation to industrial agriculture, are unstable under current social and economic conditions. The development of ecologically, socially, and economically viable forms of land use in the humid tropics will require a strategy that builds on their characteristic diversity of the humid tropics and mimics their complex ecological processes.

Research on the humid tropics is expanding as attention is drawn to their status, their role in the global environmental system, and the fate of the people who depend on them for their livelihood. Studies of sustainable agriculture must lead the way in establishing sound principles for land use and conservation in these regions. The following areas and subjects of research are suggested.

• State-of-the-art inventory, classification, and analysis of local/indigenous systems, successful experimental systems, and case studies pertaining to land-resource management. This review would serve as the foundation for, and provide insight into, elaboration of additional elements of a research agenda.

• The possibilities of restoring degraded lands and elaboration of criteria for determining when and what to restore; the possible limits to restoration; the extent to which knowledge and modern techniques are sufficient, individually or together, to restore damaged lands to functioning forests, grasslands, or farmlands. Analysis of the economic costs of land degradation and restoration should be included.

• The development and promotion of general principles and components of land management that sustain land resources under the constraints of tropical ecosystems. This research should involve careful analysis of processes (for example, the management of nutrient cycles and the manipulation of succession) that underlie the sustainability of successful systems and the identification and elaboration of new crops and innovative land-use systems that can help to overcome the short-cycle crop biases that can perpetuate degradation.

• The social forces that drive resource degradation—issues of political economy, accounting of forest goods and services, and policy. Institutional structures that mediate resource use and tenure issues should be analyzed to discover which of them promote careful resource husbandry, and under what conditions they do not. Full account must be made of the value of forest goods and services, including nontimber forest products, ecosystem services, conservation values, and costs of recuperation.

• Nutrient-cycling patterns and determination of the mass balance of nutrients and water across the full range of humid tropic agroecosystems.

• Issues of sediment additions, water quality, water availability, and water resource management. Given that tropical areas cycle more than 30 percent of the world's freshwater, encompass the largest zones of riparian vegetation, and supply many of the great fisheries of the world, these issues are particularly crucial in the tropics.

• Training U.S. and local scientists. As in other agroecosystems, scientific training is fundamental to research in the humid tropics. Any long-term strategy for improving the productive and protective capacities of tropical environments must develop the local research capacity and strengthen the local institutional support—and these need not be state or official institutions—for careful management of land and water resources. Cooperation among farmers, nongovernmental organizations, and researchers must be a key factor in elaborating new strategies and in providing extension and oversight.

SEMIARID RANGE AND SAVANNAH AGROECOSYSTEMS

Semiarid savannahs and rangelands are characterized by relatively low annual rainfall. Native vegetation—grasses and grass-like plants, shrubs, and drought-resistant trees—evolved within the limits imposed by the protracted dry seasons typical of these regions. In addition to water availability, soil acidity and inherently low soil nutrient levels act as major constraints on intensive crop production. Irrigation, where technically and economically feasible, can make semiarid lands richly productive, but careful management is required to avoid the long-term problems of salinization, waterlogging, aquifer depletion, surface water pollution, and disruption of hydrological systems.

Livestock grazing is an important economic activity in populated semiarid regions. Grazing by wild and domesticated herbivores is essential to the health of rangeland ecosystems, and traditional pastoral cultures were able to maintain human and ruminant population numbers within, and fit grazing patterns to, the carrying capacity of these lands. Human population growth, however, has placed increasing pressure on many semiarid lands. Overgrazing has increased in frequency and extent, and in some areas it has triggered the positive feedbacks that lead to environmental degradation in grassland ecosystems: decreased vegetative cover, invasion by unpalatable species, declining livestock quality, excessive wind and water erosion, soil erosion and degradation, increased susceptibility to the effects of drought, and ultimately desertification. Sustainability in semiarid regions will depend first and foremost on the recognition of the inherent fragility of semiarid lands and the tight relationships among available moisture, soil structure, soil nutrient levels, cropping and livestock patterns, the potential impacts of interventions, and human population pressures.

Semiarid range and savannah is widely, though unevenly, distributed around the world; it covers approximately 50 percent of the earth's surface. Over half of this area, however, is too cold, too dry, or too distant to support permanent concentrations of humans and their livestock and associated crops. Of the occupied rangelands, sub-Saharan Africa, from the West African Sahel through Sudan and Ethiopia to Somalia, faces the most urgent agricultural and environmental difficulties, and it will be the most important testing ground in the near future for the development of sustainable systems. Although semiarid sub-Saharan agricultural systems may have been sustainable under low-intensity exploitation, demographic, climatic, ecological, and institutional factors constitute threats to sustainability for both the short and long term.

Farmers in semiarid Africa typically grow a variety of drought-tolerant staple food crops in fields around their villages, principally sorghum and millet in pure stands or intercropped with cowpeas. Cash crops include cotton and groundnuts. Generally, households also cultivate small plots around the homestead, where they plant maize and vegetables for home use and market sale (often supplementing rainfall with water from recently drilled village tube wells). These agricultural systems are highly integrated and are frequently maintained with minimal or no external inputs. Farmers plant local crop varieties, rely on bush-fallow rotations where possible, use animal manure to maintain fertility, and use family labor to meet the highly seasonal demand for agricultural labor. Cereal yields are low, and any available marketable surpluses are primarily the result of interyear variations in rainfall. Donkeys, cattle, and small ruminants, sustained on crop residue and rangeland, are a source of manure, and in some cases of income, draft power, and food. Households typically give priority to the production of sufficient staple crops to meet the family's food needs and allocate remaining, often productive, land to cash crops and livestock.

Under any circumstances, agriculture in sub-Saharan Africa is an inherently complex, interactive, and high-risk endeavor. Sustainable approaches are badly needed because population growth has put serious pressure on the fragile natural resource base. Agricultural development in sub-Saharan Africa, as in most semiarid regions, is constrained by readily identifiable factors: water availability; soil nutrient availability, erosion, physical properties, and organic matter; the institutional and human resource base; and the policies necessary to manage soil and water resources. Where overgrazing of rangelands threatens to induce the disintegration of plant communities and soil erosion, farmers have limited resources to make capital investments, and short-term returns are necessary to make investments attractive.

Sustainability under these circumstances will depend on the capacity of the low-income, medium population density countries of sub-Saharan Africa to improve the institutional climate and develop the soil-water-crop-animal systems necessary for agricultural development. The SANREM program can make substantial contributions to this effort. The observations and recommendations that follow are based on the assumption that the sub-Saharan region offers the best returns on investments in terms of widely applicable principles and effective strategies for sustainable agriculture and resource management in semiarid regions.

Soil and water are central to sustainable agriculture across the semiarid tropics. Priority must thus be given to the development of an integrative and environmentally sound systems approach to soil, water, crop, and animal husbandry in an agroecosystem context that will sustain the natural resource base, with full consideration given to institutional and policy factors (that is, land tenure). The approach developed should not only use current knowledge in an integrative manner, but also develop new knowledge and the means for its dissemination and application. Research procedure should involve the creation of an innovative model to address this overriding priority.

In locating CRSP research projects in the semiarid zone, priority should go to those sites that can demonstrate host country institutional commitment and capacity, broad applicability of potential results to other semiarid regions, and in-country mission involvement (but not extensive mission management support; for example, the Tropical Soil Management CRSP model). Research should be conducted on a watershed level at a minimum.

Proposals for research in this agroecosystem should demonstrate the following capabilities:

• innovative approaches to system modeling that are realistic, workable, and applicable;

• integrated research experience, previous commitment to work in the semiarid zone, commitment of university cost-sharing resources, continuity of staffing, and experience in systems research and management;

• an agroecological research framework that gives full attention to biotic, abiotic, and socioeconomic factors, including analysis of indigenous natural resource management; and

• complementarity and interaction with nongovernmental and private voluntary organizations, other CRSPs, and other international agricultural research organizations.

HILL LANDS

Mountain agroecosystems constitute approximately 25 percent of the total land surface of the earth, and they contain at least 10 percent of the total population. Major mountain agroecosystems are found in the Andes of South America, throughout Central America, in the Rockies of western North America, the islands of the Caribbean and Southeast Asia, the Hindu Kush-Himalaya region of South Asia, and the mountains of East and Central Africa. In virtually all of these regions, mountains exist as large humid "islands" in an otherwise arid-to-semiarid landscape, and they serve as a source to major river systems. Although populations in the mountains are relatively low, those of the "highland-lowland interactive system" are high, and they may constitute nearly half of the total population of the earth.

It is difficult to generalize about the mountain agroecosystem, because it incorporates elements of all other ecosystems—from the humid tropic ecosystems on the eastern slopes of the Andes of South America to the arid and semiarid ecosystems of the western Himalaya in South Asia. Above all, the mountain agroecosystem must be viewed as a composite of ecosystems: a three-dimensional environmental mosaic defined by factors of altitude, slope, and aspect, and characterized by agricultural problems encountered across the full spectrum of agroecosystems. In contrast to the relative spatial uniformity of many lowland systems in which traditional agriculture has evolved, the mountain system is defined by a complex terrain that limits the availability of land suitable for agriculture, that underscores the isolation of the farmer, and that highlights the importance of terrestrial-meteorological interactions in providing the water and energy necessary for sustainable plant and animal production.

Mountain agroecosystems and adjacent lowlands are dynamically linked. Water and sediment flow from highland watersheds to lowland river basins. Human population pressures in the lowlands often force more intensive development and exploitation of upland soils, forests, and grazing lands. Although the full implications of these linkages remain subject to debate, it is clear that, at least in some cases, sustainable development of the mountain system may contribute to an increase in the nonsustainability of an adjacent lowland system—and vice versa.

As is the case with many agroecosystems, those of mountains and highlands are poorly understood. Similarly, the potential of mountain systems for sustainable use has not been determined and establishing that potential remains the fundamental challenge. This in turn requires careful assessment of the system's capacity to remain stable in response to external interventions, as well as variable natural processes from within. In the mountains, as in any agroecosystem, this assessment must be based on a thorough understanding of the complex interactions of biophysical and socioeconomic factors. Soil and water, people, institutions and cultures, and economic returns on investments of labor and capital must all be considered in the formulation of appropriate management strategies.

The development of such strategies in hill lands must be based on a dynamic model of mountain agroecosystems that can identify and evaluate alternative strategies prior to their implementation. Priority should therefore be given to the development and testing of such a model, or models. This effort could begin by making use of existing biophysical and socioeconomic concepts and data bases, which would help both to build and evaluate the model, as well as to define more clearly the gaps in existing knowledge. The mountain agroecosystem model should rely heavily on emerging computer-driven information storage, remote sensing, and data analysis technologies. Development and testing of the model should begin with the most fundamental, and potentially unstable, characteristics of the mountain agroecosystem—the soil and water "life support" resource base—and should eventually incorporate all factors, including the socioeconomic and cultural.

Once a reliable model is available, researchers can develop new techniques to evaluate factors relevant to sustainability in mountain agroecosystems, including the suitability of landscape, ecosystem, or socioeconomic units for various management options; mitigation and control methodologies; activity options and alternative agricultural technologies; comparative advantages, in biophysical and socioeconomic terms, of available methods; and the economic, production, and environmental impacts of potential interventions. Only after this phase is completed should actual site-specific development and testing of more specialized models that reflect the great diversity of mountain agroecosystems be undertaken. The ultimate objective is to develop a systems approach to the planning and management of mountain agroecosystems that farmers, resource managers, and institutions can use.

In addition to U.S. universities, this activity should involve established international institutions with demonstrated capability in research on mountain agroecosystems, such as the International Centre for Integrated Mountain Development in Nepal and the University of the Andes in Venezuela. The network of mountain scientists represented by the International Mountain Society should also be used to the extent possible.

In sum, research on mountain agroecosystems should proceed in the following manner:

• Develop and test a dynamic model of mountain agroecosystems.
• Based on that model, develop and test methodological approaches to sustainable development of mountain agroecosystems.
• Prepare training materials and opportunities, including workshops, seminars and short courses, that acquaint planners, managers, and farmers with the potentials and constraints of the mountain agroecosystem and that provide for regular local input into the development and application of the model.

INPUT-INTENSIVE AGROECOSYSTEMS

Input-intensive cropping and livestock systems are found around the world. Such systems are characterized by the application of fertilizer to maintain or build soil nutrient levels each year or each crop rotational cycle and by the use of pesticides or biocontrol methods to reduce pest losses to or below threshold levels. Input-intensive systems currently account for the lion's share of world food production. They are found mainly in lowland areas and are dominated by rice, wheat, sorghum, and corn production, particularly in countries facing heavy population pressures. The sustainability of production in such systems is a vital food security and environmental concern.

Input-intensive systems are growing in importance in temperate upland regions and in the savannahs of Africa and Latin America. In the highlands of Central America and on many islands, nearly all food is produced on sloping land, often through very intensive systems that are sustainable only if soil erosion is controlled by producing high levels of crop residues and land cover year-round. Outside irrigated regions, the rain-fed agroecosystems face a distinct mix of technical, economic, and environmental problems.

Because input-intensive systems must contribute much more prominently to total food production if world food needs are to be met, they clearly warrant increased emphasis. The SANREM program should entertain proposals from all geographic regions where input-intensive systems, as defined above, play an important role in meeting regional food needs. The top priority for research on input-intensive systems should be to assess the interactions and implications of efforts to attain higher average yields, especially as they affect long-term productivity of soil and water resources and environmental quality, both on-farm and within the region. To this end, the relationship between attainable yield goals and yield instability may be of great importance from the perspective of food security and, hence, warrant special focus in research proposals. Investigators should also be required to explain how proposed research projects will improve understanding of the roots of yield instability within the region for the crops under investigation, and of the factors that could increase sustainable yield goals.

The proposals should also emphasize the relevance of the proposed research in identifying cropping and animal system technologies that can contribute to higher average yields and improved farm income, without inordinately increasing risks or per unit production costs. Another essential component of the research proposals should be a description of any changes needed in policy, institutions, and infrastructure investments to create and sustain economic incentives and markets.

Because enhancement of the inherent capacity of soil to sustain plant growth is of critical importance in achieving sustainability, investigators should also describe how the proposed research will contribute to the design of profitable farming systems that are able, over several years, to improve (a) soil physical properties, (b) soil nutrient levels and nutrient-cycling capacity and efficiency, and (c) the ability of the soil to take in and hold available moisture without causing salinity, waterlogging, or adverse effects on off-farm water quality. Equally important, researchers should explain how the proposed research will clarify the impact of agronomic and pest control practices on below-ground soil microorganisms, the levels and virulence of plant and root pathogens, and the significance of soil fauna in nutrient cycling and water retention. Researchers should also describe how they will take into account the spatial variability in landscapes, institutions, and marketing opportunities in the design of cropping and livestock systems.

In meeting these general criteria, each proposal should include the following components:

• a description of the distinct area and agroecosystem in which the research will be conducted and the collaborative efforts that will be undertaken with local organizations and institutions;

• an explanation of the local, regional, and (if appropriate) global significance of the type of cropping systems chosen for analysis;

• a discussion, with a high degree of specificity, of the biological, ecological, physical, social, and economic conditions necessary for sustainability that the proposed research will help elucidate; and

• an evaluation of the importance of socioeconomic, infrastructure, land tenure, and policy considerations in the evolution of cropping practices that may prove unsustainable, and in the adoption of improved production methods that would evolve from successful completion of the proposed research project.

COMMON PROPERTIES AND GENERAL RESEARCH CRITERIA

Sustainability in its broadest sense will require the development of management systems that can meet changing human needs in a manner that conserves natural resources and preserves environmental integrity, especially in the various agroecological zones described above. To aid progress toward this end, the SANREM research agenda will and must vary to fit the geographical, ecological, historical, and cultural realities unique to each locale. Progress in all systems, however, benefits from the recognition that they share certain features and that comprehensive scientific understanding requires an appreciation of the similarities across agroecosystems, as well as the differences among them.

Common elements can be identified in the agroecosystems described above, and in other systems that may not fit those categories. These include physical and biological factors—nutrient cycling, biodiversity, soil and water management practices, and disturbance regimes—and socioeconomic factors—land tenure and property rights, resource policy, infrastructure, gender roles, and economic constraints. Common qualities—in particular, productivity, stability, resilience, and equity—are closely associated with health in each agroecosystem. These commonalities have direct implications for the conduct of research under the SANREM program. The new program must identify ways to focus and promote ongoing research on sustainability issues in other CRSPs; to foster the interaction of indigenous knowledge and scientific methodologies; to further the necessary integration of the disciplines involved in land use and management; and to make local participation a central element of the research process. Perhaps most important, the SANREM program must identify, select, and implement projects that can fill the gaps in current knowledge. Some of these gaps have been identified above, but others will emerge. The search for hidden factors in the sustainability formula will be an important aspect of the SANREM program, and of the systems approach it adopts.

The concentration in this discussion on terrestrial systems ought not to obscure the significance of aquatic systems, in their own right or as they relate to agricultural practices and other aspects of natural resource management. Water itself is a factor common to all ecosystems. A comprehensive scientific approach to the environment in which agriculture is practiced must account for the water resources used in, and the aquatic systems that are affected by, agriculture. As fish play an increasingly important role in the human diet (particularly in developing countries, where they often account for over 40 percent of animal protein consumed), coastal-zone harvesting and aquacultural activities must necessarily be incorporated into the sustainability research agenda. Fisheries and aquaculture entail special considerations, but they are subject to the same principles that govern the sustainability of land-based agriculture; in many regions the two are tightly coupled. Aquatic and agricultural ecosystems are also directly linked by biological and physical processes (the most broadly significant being the cycling of nutrients through waste conversion and feed and fertilizer production); by environmental concerns (especially water quality issues involving soil erosion, siltation, and the runoff of pollutants, fertilizers, and pesticides); and by the prospect of global climate change and its attendant impact on sea levels and biodiversity. These and other considerations point to the need to weigh fully the aquatic component, and its potential contribution, in research designs.

In certain agroecosystems, aquaculture may come to play a direct and highly significant role. Not all agriculture can incorporate aquaculture, but a significant proportion can, and the result can be a more sustainable production system. The entire SANREM program does not need an aquaculture component, but at least some sustainable agriculture systems should be developed with aquaculture as a functioning component of the system. The existing aquaculture CRSP can provide an excellent data base and a cadre of trained professionals able to bring their experience to the scientific exploration and development of integrated agriculture-aquaculture systems.

The control of pests is another universal feature, common to all agroecosystems, unique in its local needs, and central to the SANREM research program. In its method of investigating and responding to complex environmental phenomena in an agricultural context, integrated pest management provides a model for systems-based research and is itself a vital component of sustainable agriculture.

Underlying this identification of features common to all agroecosystems is the question the effort is meant to address: how can science best serve to inform the issues that sustainability raises? *Sustainability* is itself a relatively new term, and researchers have only begun to define the structure of the science that describes it. At this point, one can say that there are fairly well-developed principles governing agroecological systems that, if violated, make systems unsustainable; that those principles can be elaborated; that once elaborated they can be converted into hypotheses appropriate to a particular agroclimatic region; that research can be designed to validate, accept, reject, modify, or develop further those hypotheses by conducting investigations and on-farm tests in the relevant regions; and that the investigations and tests can then be evaluated and interpreted in the broader context that a systems perspective provides.

The SANREM program was created to advance this process. The specific criteria for research outlined above emphasize the needs of particular agroecosystems. Regardless of the agroecosystem under investigation, however, a successful proposal within the SANREM program will have taken into consideration the following questions:

• How does the project foster conditions and a consensus for collaboration among various constituencies?

• Are collaborative and innovative research methodologies used by the project?

• Is the project interdisciplinary?

• Does the project emphasize local and traditional expertise, knowledge, and institutional development?

• Does the project address gender issues and equity considerations?

• To what extent are intended beneficiary farmers and nongovernmental organizations integrated into the design, planning, implementation, monitoring, and evaluation of the project?

• Does the project have both applied and adaptive phases to ensure that practical results accrue for resource-poor farmers within a reasonable period of time?

• Has the project established linkages with other SANREM and non-SANREM initiatives (for example, relevant CRSPs, local or regional sustainable agriculture networks and field programs, other donor activities)?

APPENDIX E

Integrated Nutrient Management for Crop Production

Clive A. Edwards and Thurman Grove

All agricultural systems must have sources of nutrients if they are to produce crops. Prior to the discovery of inorganic fertilizers in the nineteenth century, soil fertility and nutrient supply were maintained by returning organic matter to the soil and through regular rotations and fallow periods. The work of Liebig, summarized in his book *Organic Chemistry in its Applications to Agriculture and Physiology* (1840), and the experiments of Lawes and Gilbert in the mid-1800s at Rothamstead, England, led to a progressive expansion in the use of inorganic fertilizers containing nitrogen, phosphorus, potassium, and other minor nutrients. Inorganic fertilizers enabled farmers to grow crops in much closer sequence and ultimately in monocultures, and they facilitated the separation of crop and animal production. As a result, crop production in many areas is highly dependent on inorganic sources of nutrients. Parallel to the development of inorganic fertilizers, a progressive expansion occurred in the breeding of high-yielding crop varieties that respond well to high inputs of inorganic nutrients.

The combination of inorganic fertilizers and new crop varieties has greatly increased crop yields. There has been, for instance, an almost threefold increase in crop yields in Europe and the United States since World War II, and yields have more than doubled in developing countries where the green revolution has taken place. At the same time, use of animal manures and other organic sources of nutrients has steadily decreased, which has often

Clive Edwards is professor, Department of Entomology, The Ohio State University. Thurman Grove is program officer for agroecology and environment, Winrock International Institute (formerly with the Agency for International Development).

created organic waste disposal problems for intensive animal production systems in developed countries. Decreased use of organic inputs, increased use of inorganic nutrients, and reduced rotations have ultimately led to the growing of crops such as maize in monoculture. These, in combination with heavy soil cultivations, have led to extensive wind and water erosion of poorer soils in the United States and many other parts of the world.

A considerable degree of polarization has arisen between conventional high-production farmers who depend on inorganic fertilizers for nutrient supply and those who avoid using them for what they perceive to be environmental and ethical reasons. The latter, commonly called organic or biodynamic farmers, base their crop production on organic sources of nutrients and rotations.

In many developing countries, where soils are poorly structured and low in base fertility, and where the availability of inorganic fertilizers is limited, crop production has depended on periodic clearing of the forest and cropping for only 1 to 3 years—a practice commonly known as slash-and-burn agriculture. Traditionally, this method involved cropping the area only once over the 12- to 15-year "rotation." In recent years, population pressures have reduced the interval between cropping phases, and this method has begun to fail.

THE LESSON FROM INTEGRATED PEST MANAGEMENT

The experience with arthropod pest, disease, and weed control has been remarkably similar to that of nutrient provision. Prior to World War II, pests were controlled mainly by rotations and the use of cultural techniques. The development of extremely effective insecticides, fungicides, and herbicides in the 1940s transformed agriculture and led to virtually complete dependence on pesticides. Not until the 1960s was it realized that extensive use of broad-spectrum pesticides, often applied over large areas from the air, had led to major environmental problems.

Beginning with the introduction of the concept of integrated pest management (Stern et al., 1959), there was a systematic movement toward the use of improved pesticide formulations and localized applications of minimal amounts of pesticides, combined with appropriate cultural and biological control techniques. This trend still continues, in both developed and developing countries, and it has led to significant decreases in the amount of pesticides used on many crops.

Integrated pest management holds a clear lesson for nutrient provision and management. If sustainable agriculture and natural resource management is to be promoted on a global basis, similar principles must be developed for the provision of nutrients. The use of minimal amounts of inorganic fertilizers—applied as a "topping off" only when necessary, placed in the crop row where they will contribute only to crop growth and not to weed growth, and timed for crop needs—can be combined with plant and animal organic inputs to provide an integrated nutrient management program based on principles similar to those employed in integrated pest management programs. Although agroecosystems around the world differ greatly in soil fertility, soil structure, organic matter status, and climate, the need for pest, disease, and weed control and nutrient supply are common to all systems, and the same principles can be applied to all systems to minimize off-site inputs and maximize conservation of natural resources on a global basis.

There is increasing evidence that the critical inputs for pest, disease, and weed control are the encouragement of all forms of biodiversity and the availability of organic matter. Both inputs are critical to increasing the diversity of soil organisms, which in turn are very important to providing alternative prey and hosts for insect pests and diseases, and increased competition for weeds. Increased diversity also builds up large populations of natural enemies of insect pests, diseases, and weeds, so that they are held in check by biological pressure and do not reach serious levels.

There are many ways to increase biodiversity through cultural practices, including rotations, undersowing, alley cropping, and strip and contour cropping. There are also many ways to provide organic matter, such as plowing crops and crop residues into the soil and using animal manures and a wide range of other organic wastes from industrial sources. These practices can be complemented by the use of minimal amounts of pesticides, used in optimal ways and combined into integrated management programs.

Biodiversity and availability of organic matter are also the critical factors in the availability of nutrients for crop growth. Biodiversity of crops and cropping patterns provide a broad nutrient base and promote highly active soil microflora and fauna, which can spur the breakdown of dead plant and animal materials and the release of the nutrients they contain. In particular, legumes can be the main sources of the nitrogen essential for crop growth, and rhizosphere organisms, such as vesicular arbuscular Mycorrhizae, can increase the availability of phosphorus. Supplying organic matter is the key to providing the essential nutrients a crop needs from biological sources. As is the case with pest management, the nutrient supply from natural sources can be supplemented with carefully applied inorganic fertilizers, as required, to put together an integrated nutrient management program.

The above concept of nutrient management differs greatly from the concept commonly followed—that the main useful source of crop nutrients is inorganic fertilizers. The practices involved in an integrated nutrient management program maximize biological inputs to crop production and minimize the use of inorganic amendments so as to create a much more sustainable pattern of crop production, not only ecologically and environmentally, but also economically and socially.

DEVELOPMENT OF AN INTEGRATED NUTRIENT MANAGEMENT PROGRAM

The steps needed for the practical integration of nutrient management are analogous to those used in integrated pest management. In essence, three steps are required:

1. Assess the nutrient status and needs of the system.
   • soil nutrients
   • soil type and potential for proposed cropping
   • proposed cultivations
2. Establish an economic threshold.
   • availability of inorganic amendments
   • cost of nutrient input (chemicals, manures, and labor)
   • expected yield
   • potential financial return
3. Develop a nutrient management strategy.
   • minimal inorganic chemical needs and optimal timing and placement
   • nutrient supply from plant inputs
   • nutrient supply from animal inputs
   • nutrient inputs from other potential sources of nutrients

Such an integrated nutrient management program is a critical component of the type of integrated farming systems that are essential for the development of sustainable agriculture and natural resource management (Edwards, 1989; Edwards et al., 1990).

REFERENCES

Edwards, C. A. 1989. The importance of integration in sustainable agricultural systems. Agriculture, Ecosystems, and Environment 21:25–35.
Edwards, C. A., R. Lal, P. Madden, R. H. Miller, and G. House, G., eds. 1990. Sustainable Agricultural Systems. Ankeny, Iowa: Soil and Water Conservation Society.
Liebig, J. 1840. Organic Chemistry in Its Applications to Agriculture and Physiology. London: Taylor and Walton.
Stern, V. M., R. F. Smith, R. Van den Bosch, and K. S. Hagen. 1959. The integrated control concept. Hilgardia 29:81–101.

APPENDIX F

Integrated Pest Management for Sustainability in Developing Countries

Clive A. Edwards, H. David Thurston, and Rhonda Janke

Losses of crops to pests in developing countries are extremely large. Preharvest losses are estimated at 36 percent of potential yield, and postharvest losses at 14 percent (Agency for International Development, 1990). Control of pests still depends heavily on pesticides. Unless the introduction of pests into new regions is prevented by quarantine measures or eradication, the control of imported and indigenous pests must depend on pesticides until effective pest management strategies can be developed.

Integrated pest control (IPC) and integrated pest management (IPM) were originally developed in relation to insect pest control, beginning with the publication of Stern et al.'s classic *The Integrated Control Concept* (1959). The original concept emphasized the blending of biological and chemical control measures. It was later broadened (Smith and Reynolds, 1965) to refer to "a system which uses all suitable methods in as compatible a manner as possible." This led to the further definition by the Food Agriculture Organization (1967) of IPM as "a pest management or integrated control system which in the context of the associated environment and the population dynamics of the pest species, utilizes all suitable techniques and methods in as compatible a manner as possible and maintains pest population levels below those causing economic injury." However, neither this definition nor that of Norton and Holling (1979), which stated that the aim of

Clive A. Edwards is professor of entomology at The Ohio State University, H. David Thurston is professor of plant pathology at Cornell University, and Rhonda Janke is section leader within the Agronomy Department at the Rodale Research Center.

IPM was "to develop alternative, ecologically desirable tactics for use in suppressing major pests," makes explicit the need to minimize the use of pesticides. Integrated pest management, however, encompasses more than limiting the use of pesticides when necessary to avoid economic damage (normally termed supervised control). Probably the best definition of IPM is that of the Office of Technology Assessment: "The optimization of pest control measures in an economically and ecologically sound manner accomplished by the coordinated use of multiple tactics to assure stable crop production and to maintain pest damage below the economic injury level while minimizing hazards to humans, animals, plants and the environment" (Office of Technology Assessment, 1990).

Many of the techniques examined as components of IPM—forecasting of pest attacks, development of economic injury thresholds, use of pheromones in pest monitoring, use of selective pesticides, use of resistant crop varieties, timing of crop planting, and use of appropriate cultivations and crop rotations—have already been incorporated into current pest control practices and have led to more rational use of pesticides. Although successful IPM programs have been developed for glasshouse crops and orchard fruit in developed countries, and for some field crops (for example, cotton and rice) in developing countries, adoption of truly integrated pest management programs has been relatively slow, even for insect pests.

The concept of IPM was developed originally for the control of invertebrate pests, but its principles were soon adopted successfully for the control of diseases and, later, weeds (see Table F-1). Although many definitions of IPM have been advanced, there is general agreement on the conceptual aspects. All definitions include a management approach to pest problems that involves methodological and disciplinary integration and consideration of environmental and social values. The common aim of most IPM programs is to use multiple tactics to maintain pest damage below the economic injury level and at the same time provide protection against hazards to humans, animals, plants, and the environment.

In most crop production systems or agroecosystems, the development of an IPM program involves the following steps:

1. Identify the overall pests in the system, including
   - major pests that are perennial pests and usually cause damage above the economic injury level;
   - occasional, minor, or secondary pests that cause damage above the economic injury level only occasionally;
   - potential pests that normally do not cause economic losses; and
   - migratory pests that can cause serious damage on a periodic basis.
2. Develop suitable monitoring or forecasting techniques. This involves the measurement of pest populations (numbers of eggs, larvae, insects, spores,

mycelia, nematode cysts on adults, seeds, or weeds) or amount of damage or loss.

3. Establish economic thresholds, that is, the pest population or disease incidence levels that cause losses in crop value that exceed the cost of pest management. It may be difficult to establish such levels for some weeds and diseases.

4. Develop a pest management strategy. It is necessary to identify the least hazardous chemical that can be used, with minimal dose if needed, and the appropriate cultural and biological techniques that can be integrated into a pest management strategy. The aim is to maintain pest numbers and resultant damage at economically acceptable levels with minimal use of chemicals. IPM usually targets containment rather than eradication.

5. Identify extension and outreach programs that can assist in developing and implementing a pest management strategy.

A number of pest management programs have been developed using this overall approach, particularly for pests of cotton and rice in developing countries. The programs have led to significant decreases in the use of pesticides and in such associated problems as the development of resistance to pesticides. There is an urgent need, however, to develop pest management programs that involve the integration of insect, nematode, disease, and weed management for major tropical crops. Such programs would provide the base for modifications to cover regional differences in the kind and intensity of pest attacks.

THE PRACTICE OF INTEGRATED PEST MANAGEMENT

Integrated Arthropod Pest Management

The integrated management of arthropod pests has had mixed successes. Most programs have been based on first identifying economic thresholds of damage below which control is not economically practical. Once that decision is reached, minimal amounts of pesticides are used, combined with such cultural and biological methods of control as may be available for that particular crop and region. Since pests occur in populations that are part of complex associations with those of other species, IPM must have a thorough base in ecology.

The original IPM concepts were developed for control of insects and mites. The driving force was probably the relatively high mammalian toxicity of many insecticides and fairly rapid development of cross-linked resistance to many pesticides (Edwards, 1973a,b). Most pest management programs involve some use of chemicals, and thus integrating the use of chemicals with a wide range of other techniques must be understood. Most elements

TABLE F-1 The Potential of Manipulating Chemical, Cultural, and Biological Controls for Integrated Pest Management

Current Practices	Weeds		Insects		Pathogens	
	Developed Countries	Developing Countries	Developed Countries	Developing Countries	Developed Countries	Developing Countries
Chemical inputs						
Pest threshold	a	0	+++	+	+	0
Minimum pesticide use	++	0	++	+	+	0
Forecasting	a	0	+	+	+	0
Cultural inputs						
Tillage	+	+++	a	+	a	+
Rotation	+	++	+	+++	+	++
Fallow	0	+	+	+	+	+
Cropping patterns/ intercropping	+	+++	0	++	0	+++
Mulches	+	++	0	0	0	++
Timing of practices	+	++	+	++	+	++
Flooding	0	++	0	+	0	++
Burning	a	+	+	+	+	+
"Clean" seed	+++	+	0	0	+++	+
Organic soil amendments	a	+	+	+	+	+++
Resistant crop varieties	a	a	++	+++	+++	+++
Trap crops	0	0	+	0	+	+
Green manures	a	a	0	0	+	+

TABLE F-1 Continued

Current Practices	Weeds		Insects		Pathogens	
	Developed Countries	Developing Countries	Developed Countries	Developing Countries	Developed Countries	Developing Countries
Biological control inputs						
Genetically engineered crop varieties	0	0	0	0	0	0
Genetically engineered microorganisms	0	0	a	0	0	0
Microherbicides	a	0	0	0	0	0
Allelopathy	a	a	0	0	0	0
Pest pathogens	a	a	+	a	+	+
Entomopathogenic nematodes	0	0	a	0	0	0
Pheromones	0	0	++	a	0	0
Sterile male release	0	0	+	0	0	0
Disease antagonists	0	0	0	0	+	+
Introduction of natural enemies	+	+	+	+++	0	0

NOTE: +++, Major; ++, intermediate; +, small; 0, none.

aExamples exist, but are of minor importance.

of integrated insect management can be classified under the headings of regulatory activities, biological control, or cultural control.

Regulatory Activities

Most regulatory activities are directed at preventing the introduction of pests into new areas or regions, mainly through quarantine measures. Some eradication programs, such as those for the Mediterranean fruit fly in Florida and California, have been relatively successful, but the cost in both economic and environmental terms probably precludes their implementation in most developing countries unless the program involves nonchemical methods.

Cultural Control

Before the advent of modern insecticides, manipulation of farming practices was the main pest control tool available to farmers. Some of the cultural practices available were well established and are still used extensively in developing countries. Others are relatively new and need further testing. The cultural techniques available include the following:

TILLAGE. It was widely thought that deep moldboard plowing had beneficial effects on pest insect populations, and there are some situations in which it is true, particularly for long-lived pests, such as wireworms and chafers. With the advent of conservation tillage, however, there is good evidence that some insect pest problems are decreased by cultivations, but that others are made more serious. In general, conservation tillage tends to decrease the problems associated with a range of different pests (Stinner and House, 1990).

RESISTANT CROP VARIETIES. The use of crop varieties that resist attack by arthropod pests has been a major tool in minimizing the use of pesticides and developing pest management strategies, particularly in developing countries. The international agricultural research centers have implemented large seed bank programs as the basis for developing resistant crop varieties. This process may be accelerated through genetic engineering of new strains and varieties.

ROTATIONS AND FALLOWING. The use of crop rotations has long been a major strategy in minimizing arthropod pest attacks. Such rotations are essential in controlling long-lived, soil-inhabiting insect pests effectively. More research is needed, however, to identify the rotations that effectively minimize attack by many insect pests and to define the role that fallow periods play in this approach.

CROPPING PATTERNS. It has long been understood that crop diversity decreases arthropod pest attack. Crop diversity minimizes populations of susceptible plants and maximizes the potential of natural enemies by providing alternative hosts and habitats. Much more research is needed on the effects of relatively new cropping patterns, such as intercropping, intersowing, undersowing, and amalgamation of tree growing with annual crops. There is good evidence that such cropping patterns are very effective in minimizing arthropod pest attacks.

TIMING OF FARM OPERATIONS. The attacks by many arthropods pests can often be minimized by careful study of their life cycles and the timing of farm operations, such as sowing and harvesting, to reduce pest attacks and avoid carryover of pests from crop to crop.

Biological Control

USE OF INSECT PATHOGENS. Many arthropods are attacked and killed by viral, bacterial, or fungal pathogens. A number of these, notably *Bacillus thuringiensis* and *Beauvaria bassiana,* have been developed as commercial arthropod control agents. The potential of these control agents has been reinforced by the ability to engineer them genetically to control particular groups of pests. The registration of such organisms for commercial release, however, is problematic due to anxieties concerning their environmental impact.

ENTOMOPATHOGENIC NEMATODES. Many insects and other arthropods are attacked by parasitic nematodes, of which the most important is *Neoaplectana carpocapsae* (Edwards and Oswald, 1981). There are many strains of these nematodes, all with different characteristics. But because nematode preparations can be formulated like a chemical pesticide and persist in soil for several months, they have considerable potential in arthropod pest management programs.

PHEROMONES. Many insect species possess sex attractants that attract the insects over long distances. The chemicals have been isolated, identified, and produced synthetically for many pest species. They can be used to disrupt mating or to attract insects to small areas where they can be killed chemically or by other means.

RELEASE OF STERILE MALES. A number of insect pests have been controlled successfully by rearing the male insects, sterilizing them by irradiation, and then releasing them in large numbers in the pest's territory. The sterile males mate with females to produce nonviable offspring or in some cases, no offspring. This technique was used very successfully to combat cattle screwworms in the southern United States.

Practical Examples of Insect Pest Management Programs

Cotton provides one of the best examples of a successful pest management program. Cotton is susceptible to a diverse range of pests, and a long and extensive record of heavy pesticide use is associated with its production. Wherever insecticides have been employed, a dramatic initial success has ensued for several years. In all cases, however, the number of pest species increased. This, combined with the gradual development of resistance, increased the need for insecticides, but little increase in yields has resulted. The net effect is a situation worse than the original. In Texas, for instance, the boll weevil and pink bollworm were the major pests. When heavy insecticide use was introduced, two new pests, another bollworm and tobacco budworm, developed as serious pests. Insecticide resistance soon followed. Had an effective IPM program not been developed, many farmers would have been forced out of business. The IPM program involves maintaining pest and natural enemy populations, shredding stalks and plowing in crop remnants to minimize pest overwintering, using selective insecticides timed to minimize effects on natural enemies, and using mechanical strippers, which kill bollworm larvae during harvest. Similar successes using different combinations of measures have been achieved in other parts of the world, especially, in Egypt and the Sudan.

Successful integrated insect pest management programs have also been implemented in the rice paddies of China, The Philippines, Indonesia, and Malaysia. Those programs have emphasized minimal dosing with granular insecticides, and only when economic thresholds are reached; using pest-resistant varieties; forecasting pest attack from light-trap catches and surveillance programs; timing of planting; flooding; trap cropping; using the parasite *Trichogramma;* using *Bacillus thuringiensis;* keeping ducks on rice paddies; and incorporating residues into soil. Successful integrated insect pest management programs have also been developed for maize, fruit, forest trees, brassicae, and other crops.

Integrated Disease Management

In developing countries, plant diseases have been given less attention than insect pests. As a result, integrated disease management has not had the attention it deserves. Estimates of global pest losses seldom break down the figures as to the type of pest, but in some of the most comprehensive studies of losses due to pests, the losses due to insects and diseases were similar.

Because fungicides have not caused as serious a toxicological problem for humans and wildlife as have many insecticides, less attention has been focused on their overuse or misuse. In addition, many plant pathogens cannot be controlled with chemicals, and thus pest management tactics other than the use of chemicals have been employed for decades, even during the "golden age" (1960s to 1970s) of pesticides.

Plant pathologists have usually emphasized prevention of plant diseases, rather than their eradication when they occur. Their approaches to disease management have focused on the use of chemicals only occasionally, not only because relatively few chemicals have been effective against internal plant pathogens, but also because many chemicals have not been cost-effective against pathogens, especially soil-borne pathogens. In addition, the economic return from crops such as cereals and forages has seldom been sufficient to justify the use of chemicals.

Approaches to Integrated Disease Management

Five major approaches have been used in integrated disease management efforts. First, regulatory activities are used to prevent the entry of pathogens into a crop-growing region or a crop. These include quarantines and other activities that regulate the sale and transport of infected seeds or propagating materials.

Second, host resistance, or the use of plant resistance to disease, has been a major disease management approach. Disease-resistant plant varieties developed by plant pathologists and breeders are grown on 75 percent of the land in crop production in the United States. For small grains and alfalfa, 95 to 98 percent of U.S. crops are planted with varieties resistant to at least one pathogen (National Academy of Sciences, 1968). Scientific breeding of plants for disease resistance did not begin until after the disastrous potato late blight epidemic in Ireland in 1845, as a result of which an estimated 1 million Irish people died during the ensuing famine. Traditional farmers, however, have been selecting for disease resistance for millenniums. Many of the major crops on which humans depend for food are constituted primarily of cultivars or races selected before modern agricultural science began. These races are usually genetically diverse and in balance with the environment and endemic pathogens. They are dependable and stable, and although not necessarily high yielding, they will yield a crop under all but the worst conditions. The conservation and possible use of these races in breeding schemes should be considered a priority in disease management programs.

A third approach, chemical control, is problematic, as noted, because few chemicals are available to control diseases caused by bacteria and viruses. Viruses cannot be controlled by chemicals, other than through their vectors.

Until the 1970s, almost all chemicals used to control fungi (fungicides) were broad-spectrum chemicals, applied to external plant surfaces, and they were generally ineffective against internal pathogens. Since most plant pathogens are sessile, the chemicals had to be present on plants in nearly a continuous layer before the pathogen arrived; if the pathogens did not come in contact with the fungicide where they were deposited, they could escape its effects. (Most insects, in contrast, are mobile and come in contact with a toxic insecticide even if it does not occur in a continuous layer.)   For certain diseases, however, fungicides are the only known management practice available.

Unfortunately, overuse of pesticides in traditional farming systems is common where the pesticides are available and affordable. Although traditional farmers may have considerable knowledge of their agroecosystems, their knowledge seldom includes information regarding the effectiveness of different chemical pesticides, and usually they have to rely on sources outside their traditional culture for advice. The quantity of pesticides used by traditional farmers in developing countries is still very small. The high cost of pesticides seriously limits their use in developing countries, since few farmers can afford to use them. Nonetheless, expectations regarding the effects of pesticides are often unrealistically high. For example, Rosado-May et al. (1985) interviewed 59 farmers in Tabasco, Mexico, about their management practices for the fungus disease web blight of beans (*Thanatephorus cucumeris*). Although farmers used several cultural methods of management, all of those interviewed said they were expecting a chemical solution to the problem.

Drying agents, such as ashes and chalk, for crops in storage and natural or nontoxic pesticides for control of insect vectors and pathogens are often effective in controlling disease, and their use should be encouraged wherever feasible as alternatives to toxic pesticides.

Fourth, biological control, or the destruction or reduction in populations of one organism by another, is common in natural ecosystems. Such interactions can be used in a variety of ways in agroecosystems to manage plant pathogens. Traditional farmers in developing countries have used this approach to control soil diseases, for example, through the development of suppressive soils and the use of antagonistic plants. The addition of large amounts of organic matter to soils by Chinese farmers, which often produces suppressive soils, is probably one of the oldest biological control practices. Historically, many agricultural systems have incorporated large quantities of organic matter into soil, which results in less soil-borne disease and other important agronomic benefits. This practice should be recommended whenever feasible.

The fifth approach, cultural control, or cultural practices for disease management, has traditionally been used by farmers in developing countries.

Little information is available in an easily accessible or understandable form, however, on the best cultural practices used in traditional systems for disease control. Among the cultural practices used by traditional farmers are altering of crop and plant architecture, encouragement of biological control agents, burning, adjusting crop density or depth at time of planting, planting diverse crops, fallowing, flooding, mulching, multiple cropping, planting without tillage, using organic amendments, planting in raised beds, rotation, sanitation, manipulating shade, and tillage. Most, but not all, of these practices are sustainable.

Examples of Successful Integrated Disease Management

Successful integrated management programs have been developed for a number of diseases important in modern agricultural systems. Many diseases, however, are still controlled by a single disease management practice. Maize provides an example of successful integrated disease management. Maize varieties in the United States are controlled through the use of disease-resistant varieties and sound crop management practices involving crop rotation, plowing under contaminated crop debris, and selecting optimal planting dates, planting sites, and plant populations. Chemicals play a minor role in the management of maize diseases, but seed is often treated with fungicides. For alfalfa and soybeans, *Phytophthora* root rots are managed by a combination of resistant varieties, plowing under contaminated crop debris, field drainage, and site selection. Increasingly, integrated disease management is relying on a combination of host plant resistance and cultural practices and less on various pesticides used for pathogen control.

Integrated Weed Management

Successful integrated weed management as practiced by farmers relies heavily on cultural practices that keep pressure to a minimum, combined with mulching or mechanical tillage during the first four to eight weeks of crop growth, which allows the crop to get ahead of weeds that emerge later. This head start allows crops to compete effectively with weeds, primarily through shading. The period soon after crop emergence is called the "critical weed-free period," and most weeds that emerge after this period do not affect crop yield, as determined by numerous experiments (Radosevich and Holt, 1984; Zimdahl, 1980). If weeds are a problem during this period despite control efforts, biological control agents and chemical control are options within an integrated weed management approach. Use of all cultural, biological, and chemical control measures should be considered carefully, however, for their effect not only on the target weed and the crop, but also on the environment and on invertebrates, nematodes, and microbes in the agroecosystem.

Cultural Control

Cultural control techniques to minimize weed pressure include many of the same approaches used by farmers to control invertebrate pests and pathogens. These include crop rotations to interrupt weed life cycles, fallowing, burning, flooding, plant date selection, adjusting crop density and planting pattern to shade weeds, multiple cropping, and the use of clean seed. In addition, crop varieties can be chosen that are especially competitive with weeds. Research has shown that varieties that are tall, have a high leaf area index, or rapid leaf area accumulation early in the growing season can suppress weeds better than varieties of similar yield potential but different morphology. However, crops have not been screened specifically for "resistance" to weeds in the way they are screened for insect and pathogen resistance or tolerance.

Cultural practices are considered the first line of defense against weeds, although many are abandoned by farmers who adopt a herbicide program for weed control. The cultural practices noted above should all be considered as components of a sustainable cropping system.

Mechanical Control

Tillage operations for mechanical control include primary tillage, secondary tillage, selective tillage and/or hand weeding, and tillage during a fallow period. Grazing, mowing, flame weeding, and soil solarization are other mechanical or physical options for weed control in some cropping systems.

Primary tillage turns under last season's crop residues and weed seeds. Secondary tillage, if delayed for two or more weeks after primary tillage, can destroy newly emerged weeds and create a relatively clean seed bed for the crop. Selective tillage includes rotary hoeing, cultivation between crop rows, hand hoeing, and hand weeding—all operations specifically performed for weed control. Repeated tillage during a fallow period is sometimes used to deplete the root reserves of a perennial weed. Grazing and mowing, in particular their timing, can be used for weed control in rangeland and forage crop systems. Flame weeding has been shown to be particularly effective for small-seeded crops (for example, carrots) and soil solarization, or heating through the use of a clear plastic mulch, has been used successfully for the production of high-value crops.

Any mechanical weed control practice used should be evaluated for its short-term and long-term environmental consequences. Primary tillage disrupts the life cycle of many soil organisms, in particular earthworms. Secondary tillage and selective tillage leave the soil relatively bare and loose during the first few weeks of the growing season and, thus, subject to erosion. The soil is probably no more vulnerable, however, than soil left bare through the use of herbicides. Tillage during a fallow period results in similar vulnerability, only for a longer period of time. Mowing and grazing of weeds in forage mixtures disrupts the habitat of invertebrates and microbes on the weeds or crops, although simply harvesting the crops would also cause this disruption. Timing of the disruption should be checked for its effect on the life cycle of beneficial insects in particular. Flame weeding requires the use of some fossil fuel and heats the soil surface slightly during the flaming operation. Flame-weeded crops also result in bare soil during a portion of the growing season. Soil solarization to kill weed seeds also kills or reduces the populations of microbes and invertebrates living in the upper horizons of the soil profile.

Cover Crops and Mulches

Mulch crops can be used in four ways for weed suppression, as cover crops in the rotation, live mulch crops, dead mulches, and allelopathic mulches.

Cover crops in rotation keep the soil covered and eliminate open "niches" of resource availability during which weeds can become established. A live mulch crop may be a cover crop that is allowed to remain in the field or is planted into the main production crop, which results in an intercrop, generally a relay crop system. Examples include maize strip-till planted into a clover and legumes or grasses that are intersown into maize at the final cultivation. A dead mulch system is one in which the main crop is planted into a standing mulch crop, which is then killed mechanically or chemically. Examples include the *tapado* system of Central America, in which beans are broadcast seeded into weeds, and then the weeds are mown or cut as a mulch. In the United States, winter annual cover crops, like hairy vetch, can be used for no-till corn production. Corn is slot-drilled into the hairy vetch, which is then killed by mowing, and left on the field to suppress weeds and supply nitrogen. Allelopathic mulches include cover crops in the first three categories that suppress weeds through chemical inhibition, in addition to physical effects. Well-known allelopathic cover crops in which the suppressive compounds have been identified include winter rye grain and oats.

Biological Control

Biological control of weeds through the introduction of insect pests and pathogens specific to particular species of weeds has been successful in many situations, for both annual and perennial weeds (Charudattan and Walker, 1982; Rosenthal et al., 1984). Unfortunately, these weed controls are often so specific and effective that private companies are not willing to develop or market them as control agents. Nonetheless, they hold promise for particular weed problems that may be encountered in a wide range of ecosystems, if developed and made available by a governmental or non-profit agency.

Chemical Control

Herbicides account for 42 percent of the world's pesticide sales and 49 percent of global pesticide use (kilograms/hectare) (Agency for International Development, 1990). Following the models of successful IPM insect control programs, weed researchers have been attempting to reduce herbicide use by determining economic thresholds for herbicides. The ready availability of postemergence herbicides makes this approach possible. Weed seedlings (or in some cases seeds) are counted, and based on models or previous experiments to determine the level at which yield loss will occur, herbicide treatments may be recommended. This approach has some inherent problems, however, principally involving time and labor.

The "monitor-and-spray-if-above-threshold" version of IPM can be used together with cultural and biological weed control practices to control problem weeds that are not effectively suppressed through other means—if one is confident about being able to predict when weeds are above the economic threshold, and if the appropriate herbicides are available at the right time and can be applied properly. It may be expensive to keep herbicides on hand if they are not used frequently. A major problem is the lack of ability to predict when yield loss will actually occur based on early-season weed counts. The effect of various weed species on soybean yield has been determined in the United States under controlled conditions, and those data can be used in predictive models. In the field, however, weeds occur in complex mixtures or communities and are often patchy, so even the most carefully developed model may be inaccurate under conditions that are less than uniform. Environmental conditions and soil fertility status also change the degree to which weeds and crops compete for limited resources, that is, more weeds can be tolerated without yield loss in a wet year compared with a dry year. Thus, models used to predict yield loss due to early-season weed presence should incorporate some form of weather forecast.

An integrated approach to weed management should take into account the effect of weeds and weed control practices on other components of the cropping system. The presence of weeds at certain levels may enhance insect pest control by providing habitat and/or an alternative food sources for beneficial insects. The removal of weeds through the use of herbicides, even if applied within an IPM framework, could disrupt the life cycles of beneficial insects.

An Example of Integrated Weed Control

Maize provides an example of integrated weed control. Maize, to begin with, should follow a previous crop with a different life cycle or growth habit—bush fallow in the tropics, for example, or winter wheat in a temperate climate. Mechanical means, such as burning or tillage in the tropic and moldboard or chisel plowing in temperate regions, are used to prepare the field prior to planting. These practices destroy or set back perennial weeds, and recently deposited weed seeds are burned or buried at a depth from which they cannot emerge if they germinate. Second, shallow tillage prior to planting can eliminate early-season weeds that germinate, and a delayed planting date in temperate climates allows the soil to warm up, which leads to rapid crop emergence and growth. During, and immediately after crop emergence, shallow mechanized tillage operations, such as rotary hoeing or harrowing, can be performed, followed by one or more passes with a row crop cultivator. Hand hoeing should be done at this time, supplemented by hand weeding in nonmechanized systems. For both the mechanized and nonmechanized cropping systems, timeliness of weed removal is essential at this stage.

During the rapid growth phase of maize, competition from the crop is important for effective weed suppression. In temperate, mechanized systems, high population densities combined with vigorous, high-leaf-area, high-yielding varieties are common. In tropical systems, it may be more common for lower population densities of maize to be combined with one or more intercropped food crops, (for example, beans, squash) to achieve a high leaf area index. Near the end of the maize life cycle, a cover crop can be undersown (for example, clovers or vetches in North America, *Mucuna deeringiana* and *Dolichos lablab* in the tropics) to provide competition for late-season weeds and fertility for the next crop in the rotation.

Management of Vertebrate Pests

Considerable crop losses are caused by birds and mammals, such as by rabbits on field crops and by rodents during postharvest storage of crops. Current control measure are primitive and ineffective, and attempts to control rodents in postharvest storage often lead to contaminated food. Few attempts have been made to develop any form of integrated management of such pest problems. There is an urgent need for innovative control measures.

THE INTEGRATION OF ARTHROPOD, NEMATODE, DISEASE, AND WEED MANAGEMENT

The main shortcoming in the development of IPM for many crops has been the failure to implement truly integrated control programs wherein entomologists, plant pathologists, nematologists, and weed specialists work together and with agronomists and plant breeders as appropriate. Only such an interdisciplinary effort can produce a sound integrated crop protection program that provides protection against animal pests, diseases, and weeds using all available environmentally desirable means, including manipulation of farm practices, and as little of chemical pesticides as possible.

Most agricultural scientists are trained in particular disciplines and tend to think in a disciplinary pattern. Pesticides to control arthropods, nematodes, diseases, and weeds are applied based on recommendations of entomologists, nematologists, plant pathologists, and weed scientists. Even when applied with reference to IPM principles, the methods consider the pests in isolation; little consideration is given to the effect of pesticides on other organisms in the agroecosystem. The main inputs to crop production—fertilizers, cultivations, and cropping patterns—all have major effects on biological and cultural pest control as well as the effectiveness of pesticides.

Arthropod pests, nematodes, plant pathogens, and weeds all interact strongly with each other, and their interactions must be taken into account in the planning of integrated crop protection programs. There are many examples of such interactions, some well documented, others more speculative. A number of the interactions are highlighted below.

• Insects transmit viruses, bacterial diseases, and fungi (for example, Dutch elm disease), along with other arthropods; feed on bacteria and fungi, including pathogens; attack weeds and weed seeds; and prey on nematodes and their cysts.

• Weeds can be alternative hosts for nematodes, arthropod pests, and diseases; can provide shelter for arthropod pests and their enemies; can attract or repel arthropod pests; can cause nematode cysts to hatch; can provide ground cover that favors carryover of diseases; and can be the overwintering hosts for arthropod pests.

• Pathogens can attack insects and weeds, can overwinter on weeds, and can influence the severity of other pathogens.

In addition, pesticides interact with natural pest control agents, as outlined below.

• Insecticides kill natural enemies of arthropod pests and nematodes, kill insects that feed on weeds, and kill arthropods that feed on fungi and other pathogens.

• Fungicides kill pathogens of pest insects and weeds, kill fungi that are the main natural control agents of nematodes, and kill organisms that are antagonistic to pathogens.

• Herbicides can kill arthropods and remove food of natural enemies.

All farm practices also exert an influence on the incidence of arthropod-pests, diseases, and weeds (Edwards, 1989).

• Inorganic fertilizers influence the growth of weeds, as well as crops, when broadcast; can make plants more susceptible to pathogens and increase disease incidence; can make plants more susceptible to arthropod pests; and can affect soil acidity-alkalinity, which in turn affects pathogens and the beneficial microflora in the soil.

• Organic fertilizers can decrease arthropod pest and disease incidence by increasing species diversity in favor of natural enemies, can absorb and inactivate pesticides, can provide alternative food for marginal arthropod pests, and can promote growth of fungi that control cysts and other nematodes.

• Cultivations mix pesticides into soil and bring them into contact with pests, affect the incidence of arthropod pests and diseases, increase the persistence of pesticides in soil, affect the natural enemies of arthropod pests, influence the distribution of pathogens in soils, and affect the incidence of weeds by mechanical damage by burying and bringing up weed seeds.

• Rotations decrease the incidence of arthropod pests by affecting the carryover from susceptible crops to another susceptible crop the following year, decrease the incidence of pathogens related to particular crops, minimize nematode populations, decrease weed problems, and encourage the buildup of natural enemies of arthropod pests.

• Cropping patterns provide physical barriers to movement of pests, provide an altered microclimate, and transmit diseases (Allen, 1989), and intercropping and undersowing favor natural enemies of arthropod pests, provide more competition to weeds than monocultures, and decrease attack by some pathogens.

The only way to identify the key inputs in an integrated pest, disease, and weed management program is through the development of a thorough information base and additional research on critical components. Finally, a practical model must be developed that can be tested for its effectiveness in the field. Such tests should be made in a whole farm system rather than in experimental field plots. The additional benefit of this on-farm approach is that it provides the farmer with knowledge of the relevant techniques on his or her own farm and serves as a demonstration area for neighboring farmers, thus filling an extension as well as a research role.

Integrated pest management and integrated farming systems are much more knowledge- and management-intensive than the simple use of pesticides and fertilizers on a recommended basis. They should be based on a firm understanding of the factors that exert the greatest effect on pests. Nevertheless, IPM systems can be developed progressively from a relatively simple pattern, adding components as the system and its interactions become progressively better understood. In this way the chemical inputs can be decreased progressively.

Because IPM is complex, thorough training must be provided for extension agents and farmers. For agents and farmers in developing countries, such training can be provided by bringing personnel to the United States or Europe for workshops or courses. Alternatively, it can be achieved by sending consultants to the developing countries.

THE RELATIONSHIP OF IPM TO SUSTAINABLE AGRICULTURE

The idea that pest management should be considered in the context of farm management was proposed by Vereijken et al. (1986). They suggested that a farming system consists of five main components: cultivations, fertilization, cropping patterns, crop protection, and farm economics. Central to this pattern is farm economics, which encompasses all inputs, including land, labor, buildings, machines, chemicals, and seed, balanced against yield and profits. A farming system is not just the sum of all of its components, but a complex system with intricate interactions. The concept of the central position of farm economics is in striking contrast to the perception of integrated control specialists who have assumed that plant protection, or their particular discipline, is the central component. Crop protection is only one important part of the system, and its needs and implementation depend on the system and the importance of pests.

As outlined in the previous section, crop protection measures, whether chemical or biological, all interact strongly with cultivations, fertilizers, and cropping patterns. Only in a system that minimizes chemical use, as proposed in sustainable agriculture, can undesirable ecological and environmental effects, such as pollution of soil and water by pesticides, be truly minimized. In general, integrated farming takes into account, far more than does conventional farming with standard pesticide use, the various impacts on ecosystems and society. It considers effects on (a) the quality and quantity of produce, (b) the economic viability of the system, (c) employment, public health, and the well-being of people associated with agriculture, (d) needs for energy and nonrenewable resources, (e) the quality and diversity of the landscape (clean environment), and (f) the preservation of the fauna and flora.

Currently, the conventional approach to crop production has (a) and (b) as its main objectives, and it does not take the other aspects sufficiently into account, the results of which have sometimes been undesirable or even harmful. In recent years, there have been increasing demands for a better balance among the various factors, based on a growing awareness of the problems caused by conventional, chemically based farming. Hence, the increasing need for an integrated, sustainable farming systems approach.

The Impact of Innovative Practices on Pest Management

Traditional agriculture in temperate countries has depended on deep plowing, the use of inorganic fertilizers and chemical pesticides in large fields, and growing crops in monoculture or biculture. Such practices encourage the carryover of pests, diseases, and weeds from one year to the next by minimizing overall diversity and disturbance. All of the basic components can be modified by introducing newer practices that decrease the adverse effects of pests and reduce the need for pesticides. In developing countries, the Green Revolution has encouraged a similar pattern of agriculture. In poorer soils, pest, disease, and weed control depends only on the use of cropping patterns and cultivations.

Mechanical Operations and Cultivations

Traditionally, moldboard plowing inverted the soil and buried crop residues and weeds before the preparation of a seed bed for the succeeding crop. Since the 1960s, there has been a trend toward less and shallower tillage, which has culminated in the practice of killing the previous crop with herbicides and planting the next crop directly into the plant residues. This practice requires considerable use of herbicides, but current research is examining how it can be accomplished with minimal use of herbicides. No-till farming, as it has been termed, usually involves the use of special machinery. The changes in soil displacement and disturbance, location of plant residues, and weed ecology all influence the incidence of arthropod pests and diseases. Conservation tillage leads to a completely different spectrum of weeds, with lower populations of species that need to have their seeds buried to germinate and higher populations of species that are controlled by cultivation. Similarly, some diseases and insect pests decrease in severity with less cultivations and others increase. Of 45 studies surveyed by Stinner and House (1990), which involved 51 arthropod pest species, the damage by 28 percent of the pest species increased with decreasing tillage, damage by 29 percent was not significantly affected by tillage, and damage by 43 percent decreased with decreasing tillage. Thus, tillage plays a major role in pest incidence and should be taken into account in designing farm management systems that maximize pest control.

Nutrient Supply and Fertilizer

There is good evidence that inorganic fertilizers can increase pest attack and the need for use of pesticides. When inorganic fertilizers are broadcast over a field, they promote weed growth between the crop rows and increase the need for herbicides, whereas placement of the fertilizer in the row would minimize this effect (Edwards, 1989). Inorganic fertilizers can increase the incidence of leaf diseases, such as cereal leaf disease (Jenkyn and Finney, 1981), and they can also increase the incidence of pests, such as cereal aphids. Using minimal amounts of inorganic fertilizers lessens the susceptibility of crops to pests.

Organic fertilizers, on the other hand, tend to decrease attacks by diseases (Hoitink and Fahy, 1986) by promoting the activity of fungal antagonists. They also decrease attacks by many invertebrate pests by increasing species diversity in favor of natural enemies (Altieri, 1985; Edwards, 1989), by providing alternative food for marginal pests, by promoting the activity of pest antagonists, such as fungi, that attack nematodes (Kerry, 1988) and other pests, and by building up populations of arthropod predators of pests by providing them with alternative food sources. Organic matter also facilitates cultivation for control of weeds. Thus, addition of organic matter minimizes pest problems.

Biodiversity and Cropping Patterns

In temperate countries, the trend has been to grow crops in monoculture or biculture over extended periods. Multiple cropping, that is, growing more than one crop in a single field, was common in earlier agriculture and is still the main pattern in many tropical countries. Multiple cropping systems, however, are much less common, even in developing countries, than they once were.

Multiple cropping includes traditional annual sequential cropping or crop rotations, but also such innovative practices as growing two crops in the same field in a single season; intercropping or undersowing, that is, growing two or more crops in the same field, usually in alternate rows; and strip cropping, that is, growing two crops in strips wide enough to allow independent cultivation and treatments but narrow enough to allow ecological interaction (Francis, 1986). All such multiple cropping systems increase the biodiversity of habitat structure and species, which tends to minimize the incidence of pests, diseases, and weeds (Stinner and Blair, 1989). Such innovative cropping patterns have considerable potential for incorporation into integrated, lower-chemical-input farming systems in developing countries.

Integration of Farming Practices and Pest Management

Much is known about how some agricultural practices affect pest management (Edwards, 1989). Much more information is needed, however, on how the more innovative practices interact with pest attacks so that the best of them can be adopted. In addition, the principles of IPM programs must be extended to cover whole farming systems, and in a way that involves minimal use of agrochemicals and maximum use of cultural practices. Some efforts have been made to develop simple simulation models that can provide recommendations on integrated pest management and farming systems based on simple, user friendly, question-and-answer systems (Willson et al., 1987).

Integrated sustainable farming systems of the kind proposed could have a number of important benefits. In particular, they could maximize profits by lowering expenditures on purchased chemicals such as pesticides; minimize food contamination by pesticides; decelerate the development of pest resistance to chemicals; reduce the environmental impact of pesticides on beneficial organisms and wildlife; decrease the hazards to farmers of pesticide application; and minimize soil erosion.

CONSTRAINTS TO THE ADOPTION OF INTEGRATED PEST MANAGEMENT IN DEVELOPING COUNTRIES

Education

A major constraint to the successful implementation of IPM, especially in developing countries, is a lack of knowledge about IPM at all levels. Farmers, agribusiness personnel, politicians, policymakers, the general public, researchers, extensionists, and teachers—all should be better informed about the values, strategies, and results of IPM. The level of education in developing countries varies greatly, but education on pests and pest management is generally lacking or inadequate, even though the major national activity may be agriculture and pests a limiting factor.

In developing countries, extension is usually weak and poorly supported, and extension personnel usually have very low rewards and prestige. Often, the sheer number of farmers needing service is overwhelming, and thus larger farmers are given priority. Generally, extension programs have low funding, inadequate transportation, and poorly trained personnel. They have little to offer farmers and thus have little or no impact on farming practices. IPM specialists are few in number or unknown in most developing countries. If government decision makers and policymakers could be influenced to give adequate support and training to extension personnel, and if IPM specialists could be trained and adequately supported, the benefits to the agriculture of developing countries would be economic, substantial, and long lasting.

Economic and Social Constraints

The high cost of pesticides in developing countries seriously limits their use. Few farmers can afford to use them even if they are available. In those cases where no other management tactic is available for pest management, and pesticides are needed or appropriate, their high cost and unavailability constitute serious constraints to crop production. Knowledge of IPM technology cannot be delivered without some social and economic costs as well. Even when outside agencies cooperate in the development of IPM programs in developing countries, there is considerable cost to national governments. In today's world, crushing debt and the rising costs of petroleum energy have made it difficult for many developing countries to initiate new programs and to support their existing programs. These realities make the initiation of new IPM programs and training especially difficult.

The rapid expansion in the use of synthetic organic pesticides in developed countries after World War II meant that it was possible to produce blemish-free fruits and vegetables. Regulatory agencies set strict tolerances for contaminating insect parts in processed food. These high standards are primarily for aesthetic purposes and are not essential to the production of healthful food. When such strict standards spread to developing countries, they become a major constraint to the development of IPM, especially when agricultural products are produced for export to developed countries.

Environmental Problems

Pesticides are often considered to be a rapid and efficient solution to many serious pest problems in developing countries in the short term. For the long term, however, the well-known problems of resistance to pesticides, pest resurgence, secondary pest outbreaks, environmental contamination, and toxicological problems from pesticides make their use expensive and economically and socially unacceptable. Overuse and misuse of pesticides are often serious constraints to the implementation of IPM in developing countries (Edwards, 1973a,b; Edwards et al., 1978).

Historically, many sustainable agricultural systems, such as some of those in China, depended on the incorporation of large quantities of organic matter into soil. This generally resulted in reduced soil-borne disease and nematode and insect attacks, in addition to providing other important agronomic benefits. The poor availability of organic amendments in modern agricultural systems, however, is a constraint on the improvement of overall soil fertility and the control of soil-borne plant pests and pathogens. Before World War II, most agriculture in the corn belt of the United States involved both crops and livestock. Rotations were on 3- to 6-year cycles, animal manure was applied to the soil, and rotations usually included legumes. That system has given way to cash-grain systems in which rotations, if practiced, are of short duration and most fertilizers are inorganic. The cash-grain system is not a sustainable model in the long term and is inappropriate for most developing countries. Declining use of organic amend-

ments in agriculture is a serious constraint to the control of soil-borne pests
and pathogens and to long-term sustainable agriculture, in both developed
and developing countries.

Policy Constraints

Especially during the past decade, actions by monetary authorities in
developing countries have been raising rates of interest, including those
paid by the government, to unprecedentedly high levels. The need to pay
off debts at high interest rates means that governments demand immediate
financial resources, such as taxes, from farmers, and both farmers and governments find it difficult to consider long-run, environmentally sound agricultural practices. The various lenders, such as the commercial banks, the
World Bank, and the International Monetary Fund, must re-examine their
fiscal policies if they wish to encourage sound, sustainable agriculture in
developing countries.

The governments of developing countries have been adopting domestic
farm policies, such as agricultural subsidies, that often depress and destabi-
lize world prices for many of the agricultural products from developing
countries. Unless developing countries receive a fair price for their agricultural products, they cannot afford to initiate IPM programs and educate and
train farmers and technical personnel. Thus, the policies of developing
countries are also a serious constraint to the initiation of IPM programs and
long-term sustainable agriculture.

Energy

There are many concerns today about conventional agricultural systems
that are highly energy intensive and built on a narrow genetic base, that
emphasize increasingly high yields, and that lead to monoculture and sometimes to excessive erosion, pollution, and contamination by pesticide residues. Probably the most serious of these concerns is the dependence on
fossil-fuel energy in modern agriculture. Petroleum is used to manufacture
almost all pesticides, to manufacture fertilizers, to produce agricultural machinery, to fuel the machinery and irrigation equipment, and to process and
distribute food and fiber. Petroleum is a nonrenewable, finite resource. As
the price of petroleum and its products increases, reliance on petroleum
energy becomes a serious constraint to the use of some IPM strategies,
particularly in developing countries.

Research and Extension Support

In general, the support of research and extension activities by national
governments in developing countries is minuscule. Research and extension

activities are not only poorly supported, but separate and competing. For example, only 0.26 percent of the Costa Rican national budget went to agricultural research and 0.34 percent went to agricultural extension (Stewart, 1985). Yet, the main source of foreign exchange in Costa Rica is agriculture.

REFERENCES

Agency for International Development. 1990. Reports to the United States Congress by the Agency for International Development. I. Integrated Pest Management: AID Policy and Implementation. II. Pesticide And Poisoning: A Global View. Washington, D.C.: Agency for International Development.

Allen, D. J. 1989. The influence of intercropping with cereals on disease development in legumes. Paper prepared for CIMMYT/CIAT Workshop on Research Methods for Cereal/Legume Intercropping in Eastern and Southern Africa, Centro International de Agricultura Tropical (CIAT) Regional Bean Program, Arusha, Tanzania.

Altieri, M. A. 1985. Diversification of agricultural landscapes—A vital element for pest control in sustainable agriculture. Pp. 124–136 in Sustainable Agriculture and Integrated Farming Systems, T. C. Edens, C. Fridge, and S. L. Battenfield, eds. East Lansing: Michigan State University.

Charudattan, R., and H. L. Walker. 1982. Biological Control of Weeds with Plant Pathogens. New York: John Wiley & Sons.

Edwards, C. A. 1973a. Environmental Pollution by Pesticides. New York: Plenum Press.

Edwards, C. A. 1973b. Persistent Pesticides in the Environment, 2d ed. Cleveland: CRC Press.

Edwards, C. A. 1989. The importance of integration in sustainable agricultural systems. Ecosystems and Environment 21:25–35.

Edwards, C. A., and J. Oswald. 1981. Control of soil-inhabiting arthropods with Neoaplectana carpocapsae. Proceedings 11th British Insecticides and Fungicides Conference (2):467–473.

Edwards, C. A., G. K. Veeresh, and H. R. Krueger. 1978. Pesticide Residues in the Environment in India. Bangalore, India: Raja Power Press.

Food and Agriculture Organization. 1967. Report of the First Session of the FAO Panel of. Experts on Integrated Pest Control. Rome, Italy: Food and Agriculture Organization of the United Nations.

Francis, C. A., ed. 1986. Multiple Cropping Systems. New York: Macmillan.

Hoitink, H. A. J., and P. C. Fahy. 1986. Basis for the control of soil-borne plant pathogens with composts. Annual Review of Phytopathology 24:93–114.

Jenkyn, J. F., and M. E. Finney. 1981. Fertilizers, fungicides and sowing date. Pp. 179–188 in Strategies for the Control of Cereal Diseases, J. F. Jenkyn and R. T. Plumb, eds. Oxford, England: Blackwell.

Kerry, B. 1988. Fungal parasites of crop cyst nematodes. Pp. 293–306 in Biological Interactions in Soil, C. A. Edwards et al., eds. The Hague, Netherlands: Elsevier.

National Academy of Sciences. 1968. Plant-Disease Development and Control. NAS Publication 1596. Washington, D.C.: National Academy Press.

Norton, G. A., and C. S. Holling. 1979. Proceedings of an International Conference on Pest Management, October 25–29, 1976. New York: Pergamon Press.

Office of Technology Assessment, U.S. Congress. 1990. A Plague of Locusts, Special Report. OTA-F-450. Washington, D.C.: U.S. Government Printing Office.

Radosevich, S. R., and J. S. Holt. 1984. Weed Ecology: Implications for Vegetation Management. New York: John Wiley & Sons.

Rosado-May, F. J., R. Garcia-Espinosa, and S. R. Gliessman. 1985. Impacto de los fitopatogenos del suelo al cultivo del frijol en suelos bajo differente manejo en la Chontalpa Tabasco. Revista Mexico Fitopatologia 3(2):80–90.

Rosenthal, S. S., D. M. Maddox, and K. Brunetti. 1984. Biological Methods of Weed Control. Monograph for the California Weed Conference. Fresno, Calif.: Thomson Publications.

Smith, R. F., and H. T. Reynolds. 1965. Principles, definitions and scope of integrated pest control. Proceedings of FAO Symposium on Integrated Pest Control 1:11–17.

Stern, V. M., R. F. Smith, R. van den Bosch, and K. S. Hagen. 1959. The integrated control concept. Hilgardia 29:81–101.

Stewart, R. 1985. Costa Rica and the CGIAR Centers. A Study of Their Collaboration in Agricultural Research. Consultative Group on International Agricultural Research, Study Paper No. 4. Washington, D.C.: World Bank.

Stinner, B. R., and J. M. Blair. 1989. Ecological and agronomic characteristics of innovative cropping systems. Pp. 123–140 in Sustainable Agricultural Systems, C. A. Edwards et al., eds. Ankeny, Iowa: Soil and Water Conservation Society.

Stinner, B. R., and G. J. House. 1990. Arthropods and other invertebrates in conservation tillage agriculture. Annual Review of Entomology 35:299–318.

Vereijken, P., C. A. Edwards, A. El Titi, A. Fougeroux, and M. J. Way. 1986. Management of Farming Systems for Integrated Control WPRS 1986 (IX) 2. Bulletin of the International Organization for Integrated Control.

Willson, H., F. Hall, J. Lennon, and R. Funt. 1987. Market Model: A Decision Support Program from the Computer Advisor System for Horticulture (CASH). Columbus: Ohio State University.

Zimdahl, R. L. 1980. Weed-Crop Competition: A Review. International Plant Protection Center. Corvallis: Oregon State University.

APPENDIX G

Project Bibliography

The following papers were prepared for the Forum on Sustainable Agriculture and Natural Resource Management Forum, held November 13–16, 1990, at the National Academy of Sciences facilities in Washington, D.C. This report includes adapted portions of several of these papers, and it has drawn on all of them in support of its recommendations. The Committee on Sustainable Agriculture and Natural Resource Management (SANREM) wishes to acknowledge the contributions of all the authors and to thank them for their efforts in gathering this material together on very short notice. Copies of these papers are available from the National Research Council Board on Science and Technology for International Development.

Edwards, Clive A., H. David Thurston, and Rhonda Janke. Integrated Pest Management for Sustainability in Developing Countries.

Fricke, Thomas B. Forging Effective Broad-based Linkages for Sustainable Agriculture Research Among Universities, IARCs, NGOs, and Farmers.

Gilles, Jere Lee. Social Sciences, Multi-Disciplinary Research, and Sustainability: Observations and Suggestions.

Grove, Thurman L., Clive A. Edwards, Richard R. Harwood, and Carol J. Pierce Colfer. The Role of Agroecology and Integrated Farming Systems in Agricultural Sustainability.

Hutchinson, Frederick E. International Agricultural Research and U.S. Universities.

Lal, Rattan. Soil Research for Agricultural Sustainability in the Tropics.

McCants, Charles B. Contributions of IARCs, FAO, AID and USDA to Sustainable Agriculture and Gaps in the Information Base.

Melcher, Jaye, and Pamela Stanbury. Approaches to Interdisciplinary Research: Recommendations for the SANREM CRSP, or How to Sow/Sew the Fields Together.

Yohe, John, Patricia Barnes-McConnell, Hillary Egna, John Rowntree, Jim Oxley, Roger Hanson, David Cummins, and Avanelle Kirksey. The Collaborative Research Programs (CRSPs), 1978 to 1990.

BACKGROUND DOCUMENTS AND PAPERS

In addition to the papers formally prepared for the SANREM forum, many others, covering a wide range of relevant issues, were contributed by forum participants. The committee wishes to thank those authors for their voluntary efforts and offers this list of papers and additional background material for those interested in further information on topics pertinent to the SANREM program. Copies of these papers, and of the background documents, are also available from the National Research Council Board on Science and Technology for International Development.

Agency for International Development. List and Description of the Pest Management Activities of AID's Various Missions. With cover letter.

Agency for International Development. Locust and Grasshopper Control in Africa/Asia: Programmatic Environmental Assessment.

Agency for International Development. Pest Management Guidelines.

Agency for International Development. Reports to the U.S. Congress— September 1990.

Alford, Donald. Sustainable Mountain Agriculture: A Personal Perspective.

Andrews, Keith L., and Jeffery W. Bentley. IPM and Resource-Poor Central American Farmers.

Ausher, Reuben, and Joseph Palti, eds. Establishing and Operating Plant Clinics Integrated with Extension Services in Developing Countries.

Baker, Frank H., and Ned S. Raun. The Role and Contributions of Animals in Alternative Agricultural Systems.

Barker, Kenneth R. Plant Nematology Report for AID's Sustainable Agriculture and Natural Resource Management (SANREM) Workshop.

Barnes-McConnell, Pat. Comments for the SANREM Planning Meeting.

Bentley, Jeffery W. What Farmers Don't Know Can't Help Them.

Bernsten, Richard H. Sustainable Agriculture in the Arid and Semi-Arid Sub-Saharan Africa: Reflections on Factors to Consider in Setting Research Priorities.

Bertrand, Anson R., et al. Interim Evaluation of the Integrated Pest Management Environmental Protection Project.

Committee on International Soil and Water Research and Development. Panel Meeting Draft Report. November 1990.

The Conservation Foundation. Opportunities to Assist Developing Countries in the Proper Use of Agricultural and Industrial Chemicals.

Consortium for International Crop Protection. Response to the Interim Evaluation of the CICP IPM and Environmental Protection Project.

Edwards, Clive A. Agrochemicals as Environmental Pollutants.

Edwards, Clive A. The Application of Genetic Engineering to Pest Control.

Edwards, Clive A. The Importance of Integration in Sustainable Agricultural Systems.

Edwards, Clive A., and Benjamin R. Stinner. The Use of Innovative Agricultural Practices in a Farm Systems Context for Pest Control in the 1990s.

Edwards, Clive A., Robert Hart, and John A. Dixon. A Strategy For Developing and Implementing Sustainable Agricultural Strategies in Developing Countries.

Eplee, Robert E. Integrated Management System for *Striga*.

Eplee, Robert E. *Striga:* A Review of Management Strategies.

Gallucci, Vincent F. Sustainable Yield in Coastal Zones and Aquatic Ecosystems.

Gold, Clifford S., et al. Direct and Residual Effects of Short Duration Intercrops on the Cassava Whiteflies (*Aleurotrachelus socialis* and *Trialeurodes variabilis*) in Colombia.

Goodell, Grace. Challenges to International Pest Management Research and Extension in the Third World: Do We Really Want IPM to Work?

Groenfeldt, David. Managing A Greener Revolution.

Gutierrez, A. P., B. Wermelinger, F. Schulthess, J. U. Baumgaertner, J. S. Yaninek, H. R. Herren, P. Neuenschwander, B. Lohr, W. N. O. Hammond, and C. K. Ellis. An Overview of a Systems Model of Cassava and Cassava Pests in Africa.

Hankins, Allen, and William B. Jackson. Project Evaluation: Vertebrate Pest Management Research and Development.

Hart, Robert D., and Michael W. Sands. Sustainability: Abstract Goal or Operational Objective?

Hecht, Susanna B. U.S. Development Assistance and Environment.

Herren, Hans R. Biological Control as the Primary Option in Sustainable Pest Management: The Cassava Pest Project.

Industry Council for Development. Agribusiness Development: ICB Experience and Methodology.

International Organization for Pest Resistance Management. Congress statement, 3/28/90.

International Workshop on Sustainable Land Use Systems. Recommendations and Resolutions Adopted by the Participants.

Jodha, N. S. Mountain Agriculture: The Search for Sustainability.

Jodha, N. S. Mountain Perspective and Sustainability: A Framework for Development Strategies.

Jodha, N. S. Rural Common Property Resources: Contributions and Crisis.

Jodha, N. S. Sustainable Agriculture in Fragile Resource Zones: Technological Imperatives.

Jodha, N. S. Sustainable Mountain Agriculture: Some Preconditions.

Jodha, N. S., and R. P. Singh. Crop Rotation in Traditional Farming Systems in Selected Areas of India.

Lee, John E., Jr., and Kenneth Baum. Implications of Low-Input Farming Systems for the U.S. Position in World Agriculture.

MacKenzie, David R. Evaluation of Crop Protection Research, Training, and Technology Transfer at the International Agricultural Research Centers.

Matteson, Patricia C. The Consortium for International Crop Protection: AID's Primary Resource for Advancing IPM and Pesticide Management in Developing Countries.

Matteson, Patricia C. Integrated Pest Management for Food Crops in the Sahel: AID-Funded Accomplishments, Present Status, and Proposed Activities.

McVey, James P. Aquaculture: An Important Consideration in Sustainable Agriculture.

Nepstad, Daniel C., Christopher Uhl, and Emanuel A. S. Serrao. Restoration of Degraded Ecosystems in an Amazonian Landscape.

Neuenschwander, P., W. N. O. Hammond, O. Ajuonu, A. Gado, N. Echendu, A. H. Bokonon-Ganta, R. Allomasso, and I. Okon. Biological Control of the Cassava Mealybug, *Phenacoccus manihoti*, by *Epidinocarsis lopezi* in West Africa, as Influenced by Climate and Soil.

Nickel, John L. Research for Sustainable Agricultural Development: Some Examples from CIAT.

Paschke, J. Don. Letter to Lowell Hardin, with attached summaries of two reports on the Integrated Crop Protection CRSP Planning Study.

Posner, Joshua L., and Elon Gilbert. Sustainable Agriculture and FSR Teams in Semi-Arid West Africa: A Fatal Attraction?

Rausser, Gordon C. A New Paradigm for Policy Reform and Economic Development.

Sanders, John H. Developing New Agricultural Technologies for the Sahelian Countries: The Burkina Faso Case.

Sanders, John H. Sustainability in Input-Intensive Agro-Ecosystems in Sub-Saharan Africa.

Teng, Paul S. Integrated Pest Management in Rice: An Analysis of the Status Quo with Recommendations for Action.

Teng, Paul S. Research Priorities for Diseases.

Thurston, H. David. Plant Disease Management Practices of Traditional Farmers.

Tropical Soil Management CRSP. The Role of the Soil Management CRSP in Sustainable Agriculture Production. Statement of the Board of Directors.

Van der Graaff, Nicolas A. The Role of Governments of Least Developed Countries in Integrated Pest Management.

van Huis, A., F. Meerman, and W. Takken. The Role of the University System in the Promotion of IPM in the Developing World.

Veeresh, G. K. The Potential of IPM and Its Contribution to Agricultural Stability in the Indian Sub-Continent.

Veeresh, G. K. Resource Management on a Watershed Basis—Towards Sustainable Agriculture: A Case Study.

Waite, Benjamin H., Hiram G. Larew, and A. W. Johnson. Final External Evaluation: Crop Nematode Research and Control Project.

Weil, Ray R. Defining and Using the Concept of Sustainable Agriculture.

APPENDIX H

Program Participants

The following is a list of participants in the open forum on sustainable agriculture held November 14, 1990, at the National Academy of Sciences, Washington, D.C.

Perry Adkisson, Texas A&M University
Donald Alford, Bozeman, Montana
Keith Andrews, Pan American School of Agriculture
Gerald F. Arkin, University of Georgia
Reuben Ausher, Ministry of Agriculture, Israel
John Axtell, Purdue University
Frank H. Baker, Winrock International
William Barclay, Greenpeace
Kenneth R. Barker, North Carolina State University
Patricia Barnes-McConnell, Michigan State University
Hector Barreto, Centro Internacional de Mejoramiento de Maíz y Trigo, Guatemala
Richard Berntsen, Michigan State University
Robert O. Blake, Committee on Agricultural Sustainability for Developing Countries
Lindsay Brown, IMC Fertilizer, Inc.
Harold Coble, North Carolina State University
David Coleman, University of Georgia
Carol J. Pierce Colfer, Portland, Oregon
Ousmene Coulibaly, Purdue University
William Crichton, Mill Valley, California
Pierre Crosson, Resources for the Future
Charles Curtis, Ohio State University
Charles Delp, American Phytopathological Society
Jay Dorsey, Ohio State University
Clive A. Edwards, Ohio State University

Yoseph O. El Kana, Embassy of Israel
Robert Eplee, U.S. Department of Agriculture
Monika Escher, U.S. Department of Agriculture
Richard A. Frederiksen, Texas A&M University
Margaret Freucks, House Select Committee on Hunger
Thomas B. Fricke, Marlboro, Vermont
Vincent Gallucci, University of Washington
John Gerber, University of Illinois, Urbana-Champaign
Jere L. Gilles, University of Missouri-Columbia
Grace Goodell, Johns Hopkins University
David Groenfeld, Silver Spring, Maryland
Lowell Hardin, Purdue University
William L. Hargrove, University of Georgia
Susanna Hecht, University of California-Los Angeles
James B. Henson, Washington State University
Hans R. Herren, International Institute of Tropical Agriculture, Benin
Peter Hildebrand, University of Florida
Fred Hoefner, Washington, D.C.
Diana Horne, U.S. Environmental Protection Agency
Robert Hudgens, Winrock International
Patricia Hung, U.S. Department of Agriculture
Robert Hunter, DowElanco
William B. Jackson, Bowling Green State University
Barry Jacobsen, Auburn University
Rhonda Janke, Rodale Research Center
N. S. Jodha, International Centre for Integrated Mountain Development, Nepal
William F. Johnson, Board on International Food and Agricultural Development
   and Economic Cooperation
Michael Joshua, Virginia State University
Anthony S. R. Juo, Texas A&M University
William Doral Kemper, U.S. Department of Agriculture
Kirklyn M. Kerr, Ohio State University
Agnes Kiss, The World Bank
Elizabeth Laking, Overseas Development Council
Rattan Lal, Ohio State University
Jackie Lundy, University of California-Santa Cruz
David MacKenzie, U.S. Department of Agriculture
John Malechek, Utah State University
Charles B. McCants, Raleigh, North Carolina
James McVey, National Oceanic and Atmospheric Administration
Liberat Ndaboroheye, American University
Constance L. Neely, Terraqua International
Mortimer Neufville, University of Maryland, Eastern Shore
John J. Nicholaides III, University of Illinois, Urbana-Champaign
Eliudo Omolo, International Centre for Insect Physiology and Ecology, Kenya
Robert Paarlberg, Wellesley College
Walter Parham, U.S. Office of Technical Assessment

James Parr, U.S. Department of Agriculture
Gene V. Pollard, University of the West Indies
Joshua Posner, University of Wisconsin, Madison
Edwin C. Price, Oregon State University
Albert Printz, Neill & Co., Inc.
Sunil Pulukkody, American University
John Ragland, Board on International Food and Agricultural Development and
    Economic Cooperation
Ned S. Raun, Winrock International
Thomas B. Rice, Dekalb Plant Genetics
John T. Roundtree, University of Maryland, College Park
Kerry Sachs, FINTRAC, Inc.
Ricardo J. Salvador, Iowa State University
Pedro Sanchez, North Carolina State University
John Sanders, Purdue University
Susan Schram, National Association of State Universities and Land-Grant Colleges
G. Edward Schuh, University of Minnesota
E. Adilson Serrao, Empresa Brasileira de Pesquisas Agropecuàrias, Brazil
Christopher Seubert, Development Alternatives, Inc.
El Hadji Malick Sow, American University
Jitendra P. Srivastava, The World Bank
Claire Starkey, FINTRAC, Inc.
Allen Steinhauer, University of Maryland, College Park
George Teetes, Texas A&M University
Paul Teng, International Rice Research Institute, Philippines
Legesse Tessema, American University
H. David Thurston, Cornell University
Nicolas A. Van der Graaff, Food and Agriculture Organization of the United
    Nations, Italy
Jan van Schilfgaarde, United States Department of Agriculture
Christopher Seubert, Development Alternatives, Inc.
Lori Ann Thrupp, World Resources Institute
G. K. Veeresh, University of Agricultural Sciences, India
Robert Wagner, Potash & Phosphate Institute
D. Michael Warren, Iowa State University
Raymond Weil, University of Maryland, College Park
Thomas Westing, University of Arkansas
James Worstell, Save the Children
John Yohe, University of Nebraska

AGENCY FOR INTERNATIONAL DEVELOPMENT

William Furtick, Agency Director for Food and Agriculture

Office of Agriculture
David Bathrick, Director
Kenneth Baum

Harvey Blackburn
James Bonner
Elizabeth Corter
Vince Cusumano
Tejpal Gill
Thurman L. Grove
Harvey Hortik
Chris Jones
Jaye Melcher
John Malcolm
Robert Schaffert
Allan Showler
Joyce Turk

Office of Rural Development
Eric Chetwynd
William Douglas
Terry L. Hardt
Pamela Stanbury
Michael Yates

Office of Forestry, Environment, and Natural Resources
Twig Johnson
Michael Philly

Bureau for Africa
Peter Celpert
Lance Jepson
Walter Knausenberger
Benjamin A. Stoner

Bureau for Asia and Near East
John Flynn
Alan Hurdus

Bureau for Latin America and the Caribbean
Angel Chivi
Michael Korin
Raymond W. Walderon

NATIONAL RESEARCH COUNCIL

Board on Agriculture
Charles Benbrook, Executive Director
James Tavares, Associate Executive Director
Carla Carlson, Director of Communications
Curt Meine, Staff Associate
Barbara Rice, Associate Editor

Board on Science and Technology for International Development
John Hurley, Director
Michael McD. Dow, Project Director
Jay J. Davenport, Senior Program Officer
Neal Brandes, Project Assistant
Patricia A. Harrington
Nancy Nachbar
Sunny Schlichter
Lynn Wolter

Water Science and Technology Board
Steven Parker, Director
Chris Elfring, Senior Staff Officer

Authors

JOHN AXTELL A professor of genetics at Purdue University, Axtell's major research area is plant genetics and breeding. He received his Ph.D. degree in genetics from the University of Wisconsin, and is a member of the National Academy of Sciences.

PATRICIA BARNES-MCCONNELL Since 1983 Barnes-McConnell has been the director of the Bean/Cowpea Collaborative Research Support Program where she organizes and implements the development of collaborative research between U.S. scientists and scientists from developing countries. She received her Ph.D. degree in developmental psychology from The Ohio State University. She is a member of the Board on Science and Technology for International Development.

HECTOR BARRETO As a member of the outreach staff of Centro Internacional de Mejoramiento de Maíz y Trigo, Barreto is a regional agronomist in the Central American and Caribbean program. He received a Ph.D. degree in soil science from Oklahoma State University. His primary area of research is soil fertility.

LEONARD BERRY Since 1987, Berry has been university provost and vice-president for academic affairs at Florida Atlantic University. He received his Ph.D. degree in geography from Bristol University, United Kingdom. His areas of expertise include rural water and natural resource planning and development.

BARBARA BRAMBLE An environmental lawyer, she established and is currently the director of the international programs department of the National Wildlife Federation to examine the interconnections between sustainable management of natural resources and the long-term success or failure of development in the Third World. Bramble received her juris doctor degree from George Washington University.

PIERRE CROSSON A senior fellow at Resources for the Future, he received his Ph.D. degree in economics from Columbia University. His areas of research include the economic and environmental aspects of agricultural land management including soil erosion, pesticide pollution, the preservation of agricultural lands, and climate change.

CLIVE EDWARDS For the past 6 years, Edwards has been a professor of entomology at The Ohio State University. He received his Ph.D. degree in entomology from the University of Wisconsin and his D.Sc. degree from Bristol University, United Kingdom. His areas of research include agroecosystems, sustainable agriculture, integrated pest management, and the effects of agricultural practices on soil biota.

LOWELL HARDIN (*Panel Chairman*) Hardin is emeritus professor of agricultural economics and assistant director for international programs at Purdue University. He has a Ph.D. degree in agricultural economics from Cornell University. His areas of expertise include agricultural and economic development.

RICHARD HARWOOD In August 1990, Harwood became a professor of agronomy at Michigan State University, where he received his Ph.D. degree in plant breeding. His areas of research include farming systems for small farms and organic cropping systems.

PEDRO SANCHEZ Since 1979, he has been professor of soil science and leader of the tropical soils program at North Carolina State University. He received a Ph.D. degree in soil science from Cornell University. His primary areas of research are fertility and management of tropical soils and pastures.

G. EDWARD SCHUH Since 1987 Schuh has been dean of the Hubert H. Humphrey Institute of Public Affairs at the University of Minnesota. He received his Ph.D. degree in agricultural economics from the University of Chicago. His areas of research include agriculture and food policy and economic development.

JAN VAN SCHILFGAARDE Since 1987, he has been associate director of the Northern Plains Area of the Agricultural Research Service, U.S. Department of Agriculture. He received a Ph.D. degree in agricultural engineering and soil physics from the Iowa State University of Science and Technology. His areas of research include management of water for crop production, especially by agricultural drainage.

G. K. VEERESH The director of agricultural instruction at the University of Agricultural Sciences, Bangalore, India, Veeresh is also the founder and president of the Indian Society of Soil Biology and Ecology and the president of the International Union for the Study of Social Insects. He has two doctoral degrees in agricultural sciences.

ROBERT WAGNER Wagner is president emeritus of the Potash and Phosphate Institute. He received his Ph.D. degree in agronomy from the University of Wisconsin. His areas of expertise include forest crops and pasture management and physiology.

RECENT PUBLICATIONS OF THE BOARD ON AGRICULTURE

Policy and Resources

Managing Global Genetic Resources: Forest Trees (1991), 228 pp., ISBN 0-309-04034-5.
Managing Global Genetic Resources: The U.S. National Plant Germplasm System (1991), 174 pp., ISBN 0-309-04390-5.
Investing in Research: A Proposal to Strengthen the Agricultural, Food, and Environmental System (1989), 156 pp., ISBN 0-309-04127-9.
Alternative Agriculture (1989), 464 pp., ISBN 0-309-03987-8; ISBN 0-309-03985-1 (pbk).
Understanding Agriculture: New Directions for Education (1988), 80 pp., ISBN 0-309-03936-3.
Designing Foods: Animal Product Options in the Marketplace (1988), 394 pp., ISBN 0-309-03798-0; ISBN 0-309-03795-6 (pbk).
Agricultural Biotechnology: Strategies for National Competitiveness (1987), 224 pp., ISBN 0-309-03745-X.
Regulating Pesticides in Food: The Delaney Paradox (1987), 288 pp., ISBN 0-309-03746-8.
Pesticide Resistance: Strategies and Tactics for Management (1986), 480 pp., ISBN 0-309-03627-5.
Pesticides and Groundwater Quality: Issues and Problems in Four States (1986), 136 pp., ISBN 0-309-03676-3.
Soil Conservation: Assessing the National Resources Inventory, Volume 1 (1986), 134 pp., ISBN 0-309-03649-9.
Soil Conservation: Assessing the National Resources Inventory, Volume 2 (1986), 314 pp., ISBN 0-309-03675-5.
New Directions for Biosciences Research in Agriculture: High-Reward Opportunities (1985), 122 pp., ISBN 0-309-03542-2.
Genetic Engineering of Plants: Agricultural Research Opportunities and Policy Concerns (1984), 96 pp., ISBN 0-309-03434-5.

Nutrient Requirements of Domestic Animals Series and Related Titles

Nutrient Requirements of Horses, Fifth Revised Edition (1989), 128 pp., ISBN 0-309-03989-4; diskette included.
Nutrient Requirements of Dairy Cattle, Sixth Revised Edition, Update 1989 (1989), 168 pp., ISBN 0-309-03826-X; diskette included.
Nutrient Requirements of Swine, Ninth Revised Edition (1988), 96 pp., ISBN 0-309-03779-4.
Vitamin Tolerance of Animals (1987), 105 pp., ISBN 0-309-03728-X.
Predicting Feed Intake of Food-Producing Animals (1986), 95 pp., ISBN 0-309-03695-X.
Nutrient Requirements of Cats, Revised Edition (1986), 87 pp., ISBN 0-309-03682-8.
Nutrient Requirements of Dogs, Revised Edition (1985), 79 pp., ISBN 0-309-03496-5.
Nutrient Requirements of Sheep, Sixth Revised Edition (1985), 106 pp., ISBN 0-309-03596-1.
Nutrient Requirements of Beef Cattle, Sixth Revised Edition (1984), 90 pp., ISBN 0-309-03447-7.
Nutrient Requirements of Poultry, Eighth Revised Edition (1984), 71 pp., ISBN 0-309-03486-8.

More information, additional titles (prior to 1984), and prices are available from the National Academy Press, 2101 Constitution Avenue, Washington, DC 20418 USA, (202) 334-3313 (information only); (800) 624-6242 (orders only).

RECENT PUBLICATIONS OF THE BOARD ON SCIENCE AND TECHNOLOGY FOR INTERNATIONAL DEVELOPMENT

Energy

Alcohol Fuels: Options for Developing Countries (1983), 128 pp., ISBN 0-309-04160-0.
Producer Gas: Another Fuel for Motor Transport (1983), 112 pp., ISBN 0-309-04161-9.
The Diffusion of Biomass Energy Technologies in Developing Countries (1984), 120 pp.,
    ISBN 0-309-04253-4.

Technology Options

More Water for Arid Lands: Promising Technologies and Research Opportunities (1974),
    153 pp., ISBN 0-309-04151-1.
Making Aquatic Weeds Useful: Some Perspectives for Developing Countries (1976), 175 pp.,
    ISBN 0-309-04153-X.
Priorities in Biotechnology Research for International Development: Proceedings of a
    Workshop (1982), 261 pp., ISBN 0-309-04256-9.
Fisheries Technologies for Developing Countries (1987), 167 pp., ISBN 0-309-04260-7.

Plants

Tropical Legumes: Resources for the Future (1979), 331 pp., ISBN 0-309-04154-6.
Amaranth: Modern Prospects for an Ancient Crop (1983), 81 pp., ISBN 0-309-04171-6.
Jojoba: New Crop for Arid Lands (1985), 102 pp., ISBN 0-309-04251-8.
Quality-Protein Maize (1988), 130 pp., ISBN 0-309-04262-3.
Triticale: A Promising Addition to the World's Cereal Grains (1988), 105 pp., ISBN 0-309-04263-1.
Lost Crops of the Incas (1989), 415 pp., ISBN 0-309-04264-X.
Saline Agriculture: Salt-Tolerant Plants for Developing Countries (1989), 144 pp.,
    ISBN 0-309-04266-6.

Innovations in Tropical Forestry

Sowing Forests from the Air (1981), 64 pp., ISBN 0-309-04257-7.
Firewood Crops: Shrub and Tree Species for Energy Production, Volume II (1983), 92 pp.,
    ISBN 0-309-04164-3.
Mangium and Other Fast-Growing Acacias for the Humid Tropics (1983), 63 pp.,
    ISBN 0-309-04165-1.
Calliandra: A Versatile Small Tree for the Humid Tropics (1983), 56 pp., ISBN 0-309-04166-X.
Casuarinas: Nitrogen-Fixing Trees for Adverse Sites (1983), 118 pp., ISBN 0-309-04167-8.
Leucaena: Promising Forage and Tree Crop for the Tropics (1984), 2d ed., 100 pp.,
    ISBN 0-309-04250-X.

Managing Tropical Animal Resources

The Water Buffalo: New Prospects for an Underutilized Animal (1981), 188 pp.,
    ISBN 0-309-04159-7.
Butterfly Farming in Papua New Guinea (1983), 36 pp., ISBN 0-309-04168-6.
Crocodiles as a Resource for the Tropics (1983), 60 pp., ISBN 0-309-04169-4.
Little-Known Asian Animals with a Promising Economic Future (1983), 133 pp.,
    ISBN 0-309-04170-8.

Health

Opportunities for the Control of Dracunculiasis (1983), 65 pp., ISBN 0-309-04172-4.
Manpower Needs and Career Opportunities in the Field Aspects of Vector Biology (1983),
    53 pp., ISBN 0-309-04252-6.
U.S. Capacity to Address Tropical Infectious Diseases (1987), 225 pp., ISBN 0-309-04259-3.

Resource Management

Environmental Change in the West African Sahel (1984), 96 pp., ISBN 0-309-04173-2.
Agroforestry in the West African Sahel (1984), 86 pp., ISBN 0-309-04174-0.

Additional titles and ordering information are available from: Board on Science and Technology for International Development, Publications and Information Services (HA-476E), Office of International Affairs, National Research Council, 2101 Constitution Avenue, Washington, DC 20418 USA, (202) 334-2688.